Cows, Ants, Termites, and Me

Also by Karen B. London
Treat Everyone Like a Dog: How a Dog Trainer's
World View Can Improve Your Life

**By Karen B. London
(co-authored with Patricia B. McConnell)**

Feeling Outnumbered: How to Manage and Enjoy
Your Multi-Dog Household

Feisty Fido: Help for the Leash-Reactive Dog

Way to Go! How to Housetrain a Dog of Any Age

Play Together, Stay Together: Happy and Healthy
Play Between People and Dogs

Love Has No Age Limit: Welcoming an Adopted Dog
Into Your Home

Cows, Ants, Termites, and Me

Revealing the World of Animals One Newspaper Column at a Time

Karen B. London, PhD

Copyright © 2023 by Karen B. London

All rights reserved. No part of this publication may be reproduced, distributed, or transmitted in any form or by any means, including photocopying, recording, or other electronic or mechanical methods, without the prior written permission of the publisher, except in the case of brief quotations embodied in critical reviews and certain other noncommercial uses permitted by copyright law. For permission requests, write to the publisher, addressed "Attention: Permissions Coordinator," at the address below.

Karen B. London/Animal Point Press
3100 West Shannon Drive
Flagstaff, Arizona 86001

Cover design by Julie Sullivan Brace, Shine Creative Industries, Flagstaff, Arizona

Arctic Tern photograph by Melissa Hafting/Macaulay Library at the Cornell Lab of Ornithology (ML114017081)

Paperback: ISBN 978-1-952960-02-4
E-book: ISBN 978-1-952960-03-1
Library of Congress Control Number: 2022922537

Cows, Ants, Termites, and Me: Revealing the World of Animals One Newspaper Column at a Time / Karen B. London. -- 1st ed.

To my Dad, Ralph London, whose fine example and parenting guided me in the best of ways into the person and the writer I have become

Table of Contents

Introduction 1
Author's Note 5
Watching Monkeys in Wild—Unforgettable 9
Celebrate the Illustrious Accomplishments of Insects 11
'Tis the Season for Winter Warmth 13
Seals at Home in the Water 15
Polar Bears and Problem Behavior 18
Animal Tracks Disturbing the Snow 20
Sleep Helps Birds Learn to Sing 22
Darwin's First Love Was Beetles 25
Courtship Feeding Rituals 27
Fish Can Change Sex 29
People Can Mimic Bird Vocalizations 32
The Trouble with Bugs 34
Support Your Local Monster 36
The Miracles of *Earth* 38
Puffins—A Favorite for All Ages 41
Cows, Ants, Termites, and Me 43
Sweat as a Defense Mechanism 45
Cats Like to Keep Their Dignity 47
Animals Do the Nastiest Things 50
Guinea Pigs: The Next Big Thing? 52
It's a Bird, It's a Plane, It's a What? 55
Holy Swimming Bat Rays! 57
How Salmon Find Their Way 59

Applying Animal Behavior to Real Life 61
Ants Enslave Other Ants 64
The Freshwater Alphabet Encyclopedia 66
Snow Up to Their Ears 68
Combat Fish Trouble 71
Made to Finish Your Duet 73
Blue in the Animal World 75
Condors at Grand Canyon 77
Ants Rule 80
Who Let the Cat Out of the Box? 82
Unseen Jaguars Watch, Wait 84
Rats Make Great Pets 87
Adventures in Pet Sitting 89
Tide Pool Diversity 91
Animal Movie Mistakes 94
Crows Are Smart and Social 96
The Display of the Peacock 98
Birds are Dinosaurs 101
Turkey Time 104
The Call of the Coqui 106
Where the Buffalo Roam 108
Wing Shape Affects Flight 111
Courtship Gifts 113
GloFish® a Trendy Pet 115
Why Do Cats Play with Their Food? 118
Scared of Spiders? 120
The Dance of the Honey Bees 122

Mimicry in the Animal Kingdom	124
Charging Elephants	127
Bald Eagles	129
Small but Mighty	131
Living on a Hamster Wheel	134
Snakes Are Cool and Alarming	136
What Is the Purpose of Play?	138
The Sting of Defense	141
Umwelt: The Perceptual World	143
Honeyguides, Humans Work Together	145
Creatures Great and Small	148
Don't Worry about the Reindeer	150
Unusual Mascots in Today's Rose Bowl	153
The Slow-Moving Sloth	155
Reptiles Join the Family	157
Birds, Bees, Flowers, and Robbers	160
Thoughts on Animals	162
Animals and Their Plants	165
It's Alarming!	168
Wild Beasts, Children, and Art	170
Things Moms Say About Animals	172
Domestication and the Anna Karenina Principle	175
Unexpected Insect Outbreaks	177
Food Caching and Retrieval	180
Birding by Numbers	182
Citius, Altius, Fortius	185
Swimming with a Whale Shark	187

They Look Alike, But Why?...189

Endemic, Native, Introduced, and Invasive Species............192

Echolocation by Bats..194

Getting Skunked 197

Camouflage and Hide-and-Seek...199

Schooled by Fish..201

Out of Many, None...204

Anatomy of the Lizards 206

Age-Related Division of Labor 208

Crime-Fighting Insects 211

Birds Learn from One Another 213

Not Extinct After All..215

Bioluminescence..218

Like Flies to Alcohol 220

Animal Architects...222

Social Cooking...225

That's How Dung Beetles Roll...227

Language Is Crawling with Insects 229

Power of the Mob...231

In the Beginning, There Were Termites...............................234

Frogs of My Childhood..236

Biting, Spitting, Leggy Marvels...238

The American Crocodile 241

Crowds, Cues, and Wildlife 243

Flying Animals That Can't Fly 245

Birds Make Our Spirits Soar..248

Sleeping Beauty...250

The Significance of Sheep 252
Can You Hear Me Now? 255
Swifter, Higher, Stronger 257
Sea Devil Romance 259
Temperatures Troubling to Turtles, Too 262
Island Biogeography 264
Elephant Moms—And All Moms—Deserve Applause 267
The Sting of Colony Collapse Disorder 269
Animal Fathers 271
Handling the Desert's Dryness 273
Surprisingly Related 276
A Worm of Legend 278
In Defense of Poop 280
Fun with Names 283
Large AND Loud 285
Black Cats: Lucky or Unlucky? 287
Who Bit Me? 289
Life on the Farm 292
Reindeer Are the Chosen Ones 294
So Many Species 296
Egyptian Animal Mummies 298
Identifying Individual Animals 301
Our Deadly Violent Relatives 303
Symmetry, Beauty, and Mating Success 306
The Birds and the Bees . . . And the Pollen 308
Elusive Giants 311
Parasitoids Influence Host Behavior 313

Mothers Are Givers 315
Camel Envy 318
You Are What You Eat 320
Fatherhood Lessons from the Red Fox 322
Ocean Love 325
Animals Use Tools, Too 327
Animals Can't Do What? 329
What's Up with Giraffe Necks? 331
America's Speed Demon 334
Sponges Are Animals 336
Biomimicry: Nature's Solutions 339
Blue Whales Feast Every Day 341
Insects of the Snow 343
Tern the Table on Santa's Reindeer 346
Acknowledgments 349
About the Author 353

Introduction

My professional life revolves around dogs—dog training, dog behavior problems, dog advice columns, dog blogs, dog books, and, of course, dog hair. (It's beyond me how lint rollers fail to qualify as deductible business expenses. Who's with me?) More people know of me through my work and writing about our best friend, the domestic dog, than anything else, but I didn't come to a career with dogs directly. I arrived there through a love of all kinds of animals, and specifically through a fascination with their behavior.

As a child, I did love dogs, but I was also incredibly fond of frogs, hamsters, whales, sea anemones, earthworms, cats, butterflies, fish, seals, lizards, rabbits, sea stars, sea urchins, bears, barnacles, and those delightful little under-the-rock animals that people typically call either roly-polies or pill bugs. I spent a lot of time observing and enjoying animals of all kinds. In the interest of being totally honest with you, it has to be shared that as a kid I was terrified of spiders. Over the years, I've come to recognize the error of my ways and now appreciate how amazing they are. Please don't judge me for taking such a long time to come around.

It can hardly have surprised anyone who knew me that I majored in Biology as an undergraduate at UCLA, or that I decided to go to graduate school in Zoology. It might have seemed a bit curious to some people that I moved to Madison, Wisconsin to study wasps, and even I had my moments of dismay over the development. Between undergrad and grad school, I spent a little less than 6 months teaching marine biology and island ecology to 11 to 17-year old students on Catalina Island off the coast of Southern California, so the move to the Midwest was a bit of a shock. On one cold day as I walked home shivering, I distinctly remember thinking, "Oh my heavens, how did this happen? Did I really leave the island to come to Wisconsin to study social insects of all things?"

Indeed I did, and it was one of the best decisions of my life. I spent six field seasons in the tropics—three in Costa Rica, two in Venezuela, and one in Mexico, and I loved it! Besides my dissertation research, which focused on neotropical social wasps, I studied ants and birds a bit, too. On top of that, I spent time watching and sometimes interacting with the most amazing species—anacondas and eyelash pit vipers, ocelots, storks, army ants, howler monkeys, katydids, toucans, peanut-headed bugs, and so many more.

Telling stories about the animals I've seen while scuba diving, having tropical adventures, and conducting field biology is a great joy in my life. When I was offered the opportunity to take over the animal column in my local paper, the *Arizona Daily Sun*, I began to write those stories down, and also to write about animal research and discoveries that did not involve me except as a reader of research by other people.

After more than a dozen years writing this column about a great variety of animals, usually emphasizing animal behavior, I've decided it is time to compile the first volume of them. Inside, you will find 145 columns from December 2008 through December 2015—the first 7 years. My hope is that these collected columns will both entertain and educate—as animals have always done for me.

Author's Note

During the years I have been writing about animals for the *Arizona Daily Sun*, there have been multiple editors in charge of this column, and also changes in my own approach to various grammatical details. So, if you love consistency, please accept my apologies because you won't find it in these pages. My relationship with the Oxford comma (I *love* it, but that has not been uniformly true of the editors) has called for flexibility. As the years have gone by, I have changed the pronouns I use for animals along with much of the rest of society. I used to mix in he/she/it, and now I am much more inclined to use "they" for individuals as well as for groups. I hope you can think of the evolution in my use of gender pronouns as a mirror of society over time and embrace the changes!

There's no single explanation for the variability in capitalization you will see within this book. I generally do not capitalize the common names of species, but I make an exception for birds, as is the convention for many scientists who study them. (Because so many bird names are descriptive, it can be tricky without cues from capitalization to be clear. If someone writes that they saw a yellow warbler, did they see an individual of one of the many warbler species that are yellow in color or did they see a member of the Yellow Warbler species?) Of course, there are instances where I did capitalize common names of other types of animals, so don't seek consistency there, either. My attempts to change those for this book are bound to have let a few slip through.

And the inconsistencies don't stop there! There are issues with my irregular use of who/that/which when referring to individual animals and to species. The proper use of these relative pronouns is not something that I allowed to confine me, so let freedom ring and all of that. I use an absurd mix of metric and imperial measurements. (I make no excuses for that except to say that I am a person born and raised in the United States with scientific training, so I am comfortable with both and therefore use them both.) I sometimes spell out numbers and sometimes use numerals, and oh, the list of inconsistencies goes on.

Instead of editing the columns for uniformity, my plan was to compile them as they appeared, only making the changes that my wonderful editor, Eileen Anderson, convinced me were essential. She will not be alone in thinking perhaps more should have been made, but I took a minimalist approach. So, I have generally left the columns as they originally appeared, but not completely. Few changes were made except a small number of grammatical things that bugged me, which were only a small percentage of the ones that Eileen found less than stellar. There, that should clear everything right up, eh?

Feel free to consider these issues (as I do) a writing parallel for the complexities, inconsistencies and quirks to be found in the animal kingdom! Hopefully you can now enjoy the book without finding the inconsistencies troubling or confusing. They are just an analogy, a meta approach if you will, to the wildness of the animal world that inspired this book in the first place.

Watching Monkeys in Wild— Unforgettable

December 8, 2008

Should one ever have the opportunity to spend time in the hot and humid land near the equator, there are certain tropical experiences worth pursuing: drinking milk directly from a coconut, swinging from a vine, accidentally falling asleep in a hammock, watching and listening as hundreds of parrots fly overhead, attempting to observe the green flash as the sun falls below the horizon, sampling a fruit you've never heard of, and seeing monkeys in the wild. It is the last of these adventures that particularly captivated me during my most recent trip to the tropics, which took me to Ometepe Island in the middle of Lake Nicaragua.

Having spent so much time in the tropics, I have been awakened hundreds of times by the loud, deep, and sometimes alarmingly close sound of male mantled howler monkeys (*Alouatta palliata*) calling back and forth to one another with vocalizations that can be heard a full kilometer away, even through the dense tropical forest. While I have grown tired of waking up with the sun day after day in that manner, I never fail to relish each opportunity to watch howler monkeys in the trees.

One observation was especially memorable because I was with a group of students from Northern Arizona University in the Tropical Forest Insect Ecology class, and for many of them, this trip to Nicaragua was their first visit to the tropics, and therefore their first opportunity to see wild monkeys. On this particular day, we saw a female monkey with a very new baby, and we were able to watch them for the better part of an hour. We can only guess the baby's age, but it looked to be no more than a couple days old, based on its size.

Seeing a mother with her brand-new baby cradled against her body made her seem more like us than in any other sighting of monkeys I have ever had. Her warm brown eyes observed us as we observed her, and we wondered so many things. How recently had she given birth to this baby? Why wasn't she with the rest of her troop? Was she too slow to keep up so soon after giving birth? Did she choose to be alone during the birth? Was the location of the troop in an area that she considered unsafe for her new baby? How did she feel about having so many humans staring at her, chattering incessantly as most primates, including our own species, tend to do? None of us knew the answers to these questions, but as scientists, we found all of them interesting fodder for discussion. Even more importantly, we shared an experience that we will always remember.

For anyone who has a list of things to do during the limited time provided by one lifetime on this planet, consider adding an item about seeing monkeys in the wild. Observing another primate in its natural habitat connects us to nature and activates our love of all living creatures as few other encounters can.

This was my first column for *The London Zoo*, which has no association whatsoever with the actual London Zoo and is merely a reference to my last name and the animals that are the subject of the column. It's easy to internalize that our fellow primates are close relatives when seeing them in their natural habitat acting in ways that are so familiar.

Celebrate the Illustrious Accomplishments of Insects

December 15, 2008

Millions of years before an engineer took pencil to paper to design a bridge, argued over funding, ran over schedule, and got criticized on aesthetic grounds, ants were building their own bridges. Old world driver ants and neotropical army ants use their bodies as building blocks to form bridges that are quite complicated in structure and that enable other members of their colony to cross over small streams or gaps between roots and vines.

There are physical triumphs beyond bridges that insects accomplished long before any *Homo sapiens* even began to ponder doing them. Both honey bees and termites have air conditioning systems to cool down their nests. Bees use forced evaporative cooling by having a few individuals stationed at the entrance to the hive who ventilate the hive by constant fanning of their wings.

Some African termites have a more complex system that requires a particular form of nest architecture. Their nests can be up to 5 meters tall with a diameter at the base of 5 meters. The air in the center of the nest heats up with the breathing and activity of millions of individuals. As this air heats up, it rises in the nest to the top region, and then travels down tubes built into the walls of the nests. As this air moves, it pulls in fresh, cooler air, and due to the high humidity within the nest, evaporative cooling also lowers the nest temperature.

Besides using physical principles to control temperature, insects make use of chemical properties for defense. For example, bombardier beetles respond to being disturbed by mixing chemicals they store within their body that, when so mixed, produce a spray of chemicals that can reach the temperature of boiling water and thus deter their attackers. Even more amazing are the ground beetles that have a defense something like Mace. If a male is bothering a female by trying to mate with her, she can incapacitate him with a chemical defensive spray and then escape.

Along other lines, insects have been in agriculture far longer than humans have. For example, there are insects that grow their own food by gardening. Best known of these are the leafcutter ants of Central and South America. These ants collect pieces of leaves and flowers that are used as a substrate to grow fungus. The ants do not eat the leaves, but rather they eat the fruiting bodies of the fungus they cultivate in their fungus gardens. Another form of agriculture practiced by ants is the tending of domesticated animals. There are ants that protect aphids and provide transportation to the aphids' host plants in return for food, which consists of honeydew produced by the aphids.

It is a surprise to many people that we have so much in common with insects, but what is perhaps most startling is that so much of what we share was going on in their world long before it was happening in ours. If imitation is the sincerest form of flattery, surely we are paying a multitude of compliments to our insect neighbors.

The fascination I have with all bridges, including ant bridges, is natural with the last name of London!

'Tis the Season for Winter Warmth
December 22, 2008

Ah, traditions. Nobody can deny how much they contribute to making this festive season special. With winter officially here, it's time for my annual tradition of being astounded that so many animals survive months of extreme cold. And they do it without a wood stove, hot cocoa, or wool socks! (I shiver when it drops below 70 degrees, so I am pretty easily impressed with tales of winter survival.)

It's amazing that animals no bigger than a human foot can survive months of freezing temperatures. The range of adaptations they have for cold tolerance is extensive. There's even a whole category of animals called "snackers and nappers" who survive the winter with a dedicated approach to eating and sleeping.

Hibernation means being in a dormant, sleep-like state throughout the winter and living off the body's fat reserves while having a decrease in body temperature, heart rate, and metabolism. The approach to surviving the cold season employed by snackers and nappers is different than the strategy of true hibernators. They are not in a sleep-like state for the whole winter. Instead, it's only in the coldest times that they enter a state of torpor, during which they do survive by using their bodies' own fat reserves. Throughout the winter, whenever the weather gets a bit better with warmer temperatures, they rouse themselves and forage for food.

Two of the most common animals in this category are the chipmunks and squirrels. Chipmunks tend to make burrows under trees. They are able to stay warm by creating a nest of leaves, grasses, and other plant materials. They wake up to snack, and mainly eat the food that they have stored in and around their dens. Their large cheek pouches allow them to carry a lot of food to their den during the late summer, when they are very busy storing nuts and seeds.

In contrast to chipmunks, squirrels make their cozy winter retreats in hollow trees. These winter dens, called dreys, are constructed using a combination of twigs, leaves, and grass. Sometimes a squirrel will take over an old bird's nest and make it fit for winter by using these same materials to raise the walls and add a roof. Besides the protection of the nest, squirrels are able to stay warm because they grow a thick coat for the winter and wrap their bushy tails around themselves for extra insulation.

On some cold days, squirrels remain in their nests, feeding on stored food. When squirrels wake up to forage, they use their sense of smell to locate nuts and pine cones that they cached in the fall, but they also spend a considerable amount of time exploiting the enormous quantity of food available at birdfeeders.

This year, as I struggle with winter's harshness, I plan to take my cue from these rodents. It's always tough to get into a new routine, but I will start by foraging for some chocolate and taking a short siesta. I expect to be warm all winter, and come spring, be bright-eyed and bushy-tailed.

As a member of a species so sensitive to cold (a tropical species, really!), it is hard to comprehend the abilities of so many other animals to survive the winter.

Seals at Home in the Water

January 5, 2009

The seal swam by so fast that my initial reaction was to mistake it for a torpedo. This was odd because as a biologist, I am generally more inclined to think of living organisms than weapons. However, it was the first time I had seen a seal while scuba diving, and the ease with which the seal moved through the water was an entertainingly stark contrast with my own experience. The difference was so huge that the seal seemed too fast to be alive. It moved so gracefully and quickly, that in one second, I gained a new understanding of how poorly suited humans are for the water. The seal's streamlined body form and physiology are marvels of adaptation to life in the sea, and people just don't have the same advantages.

On land before a dive, scuba fins force humans to walk backward, and masks allow us to see only forward. Our eyes get red from the salt water, we get giant suction marks from the mask, and it is extremely likely that when we surface, what was inside our noses will be on our faces. Very attractive. Even the really good swimmers among us move and turn slowly in the water. We wear a wetsuit, weight belt, snorkel, buoyancy compensator, tank, regulator, hood and gloves, among other gear, just to be underwater for approximately the same amount of time that it took to suit up.

In contrast, seals can easily stay underwater for 30 minutes (and sometimes up to 2 hours), which they accomplish with a variety of physiological adaptations to life in the water. Their heart rate slows during dives from about 100 beats per minute to about 4-6 beats per minute. They shunt blood away from parts of the body that can tolerate low oxygen levels, and the blood goes mostly to the heart, lungs, brain, and other vital organs. They store oxygen in their blood, not their lungs, allowing greater quantities of oxygen to be stored. Organisms that have high quantities of the oxygen-binding protein myoglobin in their muscle cells are able to hold their breath longer, and seals have more than 10 times as much myoglobin in their muscles as humans do.

While underwater for extended periods, the streamlined shape of the seal's body allows rapid acceleration through the water, and the fins provide both power and the ability to change directions with great speed and agility. Though they typically move at much slower speeds, seals can swim up to 19 kilometers per hour (12 miles per hour).

For comparison, when Michael Phelps swam a world record of 1 minute, 42.96 seconds in the 200-meter freestyle at the 2008 Beijing Olympics, he was going 7 kilometers per hour (about 4.3 miles per hour). He sustained this speed for less than 2 minutes, was aided by his initial dive plus three turns and is considered to have humanity's all-time most perfect build for moving quickly through the water. He is an amazing swimmer—except when he has the misfortune to be compared to the average seal.

Even though seals can easily swim about three times as fast as Michael Phelps, I admire him for being as seal-like as any human ever—and that is a huge compliment.

Polar Bears and Problem Behavior
January 12, 2009

It is understandable if you've recently taken a short break from shoveling snow, contemplated how very much our Arizona high country resembles the arctic, and scanned the landscape hoping to see a polar bear. I've only seen polar bears in zoos, and regrettably, many were not doing well psychologically. Polar bears are prone to develop behavioral abnormalities in captivity.

Many zoo animals exhibit a repetitive behavior, or stereotypy, that is performed for long periods of time. The most common ones are pacing back and forth, excessive grooming such as pulling out their own fur or feathers, repetitive head movements, walking in circles, or swaying their body from side to side.

In zoo animals, stereotypies often stem from lack of stimulation, excessive stress, or frustration at not being able to perform certain types of behavior such as hunting, exploring, mating, or escaping other animals. Many animals, especially predators such as polar bears, have brains that developed to solve complex problems and to make hundreds of decisions each day that are critical for survival. Put them in a zoo where their daily life is drastically different than in the wild, and their behavior can frequently succumb to the abnormal. The most common stereotypy exhibited by captive bears worldwide is pacing back and forth. These bouts of pacing can be so consistent that each "lap" that the polar bear takes has the same number of steps. Many captive polar bears pace in this manner for hours every day. Larger groups seem to exhibit fewer of these repetitive actions, and the more females that are in a social group, the lower the incidence of them.

Many attempts to lessen stereotypies in polar bears involve the environmental enrichment strategies that have succeeded in helping other species. General enrichment to stimulate the bears cognitively by providing toys or making food harder to acquire by hiding it or freezing it has had limited and mixed success. The general consensus among polar bear specialists is that these typical forms of enrichment are not overly successful in lessening the frequency of polar bears' repetitive behaviors.

One successful treatment is allowing polar bears constant access to both the exhibit area and a holding area so that they can choose where to be. Also, several studies found large declines in repetitive actions after redesigning exhibits. It is unclear whether the bears behavior improved because of the novelty of a new living space or as a result of the specific changes, which included deeper pools, boulders in the water, places to get out of eyesight of other bears, new water filtration systems, making a rough beach area smoother, turning off a loud waterfall, and using more natural materials in the enclosures.

Maybe polar bears in zoos really need more of what we all need: wide open spaces, freedom to move around, mental and physical challenges, and a good social life.

More research is needed to explore all of these possibilities further in order to improve the lives of captive polar bears. On the subject of recurring human actions, there is no effective treatment for the repetitive shoveling of snow.

Several feet of snow to shovel during the first few weeks of winter that year made it difficult to think of anything else but snow.

Animal Tracks Disturbing the Snow

January 19, 2009

While out sledding this weekend, our 5-year-old son soon tired of whizzing down the hill at increasingly high speeds and began to follow the animal tracks that were all around. Besides our own tracks, we saw marks made by rabbits, squirrels, deer, and dogs.

Although squirrel and rabbit tracks are similar, telling them apart is straightforward. Both animals leave back paw prints that are side-by-side, but only the squirrel's front paws land this way. Rabbits' front paws land one in front of the other.

We followed tracks all over the hill, winding around the trees, and noticed that many of the trails we saw were actually connected, and made by the same individual. It was clear that some of these animals were traveling pretty long distances. (It's probably easier to cover some serious ground in the snow when you are not sinking up past your knee with every step, as I was.)

Beyond that they sometimes travel very far, there's a lot you can tell about animals' behavior from their tracks. Most obvious is the direction of travel, although even determining that depends on the clarity of the tracks. The distance between prints indicates speed of travel, with tracks spaced farther apart made by animals traveling faster. Faster animals make blurrier tracks and leave more snow piled around the front edge. Tracks that meander are usually made by animals searching for something in an unknown location such as food, whereas tracks in a straight line tend to be made by individuals who have a purpose, such as returning to a den, chasing prey, or escaping a predator.

After following one set of meandering squirrel tracks for some time, we noticed that the tracks became much further apart and less clear over the space of several meters, and that the tracks were now in a straight line rather than a wandering trail. Finally, the tracks ended at a point where the snow was roughed up, and we could see squirrel droppings there. Next to this spot was what looked like the impression of feathers in the snow. Detectives we are not, but it was not hard to piece together a likely scenario, which appeared to involve a squirrel that had walked upon the snow for the last time.

We imagined that a squirrel was moving along when it noticed a predatory bird in the vicinity and closing in. While unsuccessfully attempting to get to safety, the squirrel ran faster and therefore left tracks spaced further apart until the bird got it. The area of disheveled snow and droppings was presumably where the scuffle occurred when the bird captured the squirrel, at some point allowing its own body to contact the snow and leave an impression of feathers in it.

This recently departed squirrel and the predatory bird that nabbed it left behind evidence of what happened for us to understand without our having seen the animals themselves. It's like the difference between a clever thriller and a slasher movie. Sometimes the suggestion of a dramatic event is more intriguing than actually seeing it.

Observing animal behavior directly is my favorite part of being an ethologist, but it is also fascinating to surmise what happened from signs left by the animals who are no longer present.

Sleep Helps Birds Learn to Sing

February 2, 2009

Sleep is a many splendored thing.

Many parents of newborns can talk of little else and even daydream about it uncontrollably. The more researchers investigate sleep, the more we realize how amazing it is. New studies continue to reveal how vital it is to the proper functioning of the brain as well as to learning and memory in many species.

Sleep is critical for the nighttime consolidation of song learning that takes place during the day in birds, according to biologists Sylvan S. Shank and Daniel Margoliash. Their research appeared in the article "Sleep and sensorimotor integration during early vocal learning in a songbird" in a recent issue of the journal *Nature*. When juvenile Zebra Finches listened to adult birdsong during the day and also practiced singing, their brains showed an increase in certain types of neuronal activity during that night's sleep, and these changes led to improvements in the birds' singing the next day.

While it has long been known that sleep influences learning and memory, this is the first study to make direct observation of nighttime brain activity related to song learning and also the first to investigate the brain changes that occur at the onset of song learning by birds. Their brains' activities reflected the specific song the birds heard during the day. Birds that heard different songs also exhibited different activity in their brains at night. These differences suggest that while they slept, the young birds "practiced" the specific song they heard during the day.

In the areas of the brain related to song learning, the changes that occurred overnight took place before any actual changes in singing were observed. That is, birds' songs did not get better during the day as they practiced singing, but rather showed improvement the next day after the bursts of brain activity that had occurred during sleep. The authors suggest that the reactivation of sensory information in the brain during sleep may be an important way that new skills are learned and improved.

The improved singing that the birds exhibited on the day following the changes in nighttime brain activity seems to be induced by the combination of hearing adult bird song and listening to themselves practice singing. In order for the bursts of neurons firing to occur in the brain at night with the subsequent improvement of singing the next day, the birds had to both hear an adult bird song and be able to hear themselves sing. If the birds were prevented from getting auditory feedback from their own singing by the playing of white noise, neither the characteristic change in brain activity at night nor the subsequent improvement in singing the next day were observed.

The take-home message from this study is twofold: First, sleep is critical for the learning of bird song. Second, I can now consider blaming my generally horrendous singing on my lack of sleep. Perhaps to prevent people who sing as poorly as I do from ending up on TV, only those who pass a fatigue test should be allowed to audition for *American Idol.*

Consolidation of learning in multiple species continues to be an area of interest for researchers, with both sleep and play being useful behaviors after learning something new.

Darwin's First Love Was Beetles

February 9, 2009

Feb. 12, 2009, marks the 200th anniversary of Charles Darwin's birth. His scientific contributions include writings on the classification of barnacles, how earthworms shape the landscape, animal domestication, finch diversity, and insects' role in the cross-pollination of orchids.

Darwin's profound insights into the natural world include how atolls form, the elucidation of universal expressions of emotion in animals, and of course, there's his master work, *On the Origin of Species*. In this, his best-known work, Darwin presents natural selection as a mechanism for the process of evolution.

Less well-known than these major studies is that Darwin's first love in the biological world was beetles. He was so enamored of these insects that he collected them during much of his free time, and also during a considerable amount of the time that he should have spent studying medicine and later theology.

In one famous account, he found three beetles in quick succession, each of which would have enhanced his personal collection. Having grasped one in each hand, he had no way to collect another without losing one of the first two. In his eagerness to have all three, he popped one into his mouth expecting to grab the third with his now-free hand. Regrettably, the beetle in his mouth gave off a noxious chemical, causing him to spit it out, and miss his opportunity to collect the third.

Overall, it was not a satisfying collecting expedition, but it illustrates his passion for beetles.

Darwin chose a successful group of animals to adore. One in five known animal species is a beetle, which is the source of the humor behind biologist J.B.S. Haldane's oft-quoted (and possibly apocryphal) comment that a study of the natural reveals in the Creator "an inordinate fondness for beetles."

One indication of beetles' success is their diversity. Popular insects such as fireflies, ladybugs, and the dung rollers that were revered in ancient Egypt are all beetles. Also in this group are soldier beetles, tiger beetles, long-horned beetles, tortoise beetles, jewel beetles, screech beetles, rhinoceros beetles, stag beetles, bombardier beetles, whirligig beetles, water boatmen, backswimmers, click beetles, giraffe beetles, violin beetles, deathwatch beetles, fire-colored beetles, and the most delightfully named handsome fungus beetles.

Beetles contain the heaviest insects. Specimens of Goliath beetles are true behemoths, weighing in at 3.5 ounces (100 grams). That may not seem large until you put that weight into perspective: consider how unpleasant it would be to have a colony of Goliath beetles show up at your picnic instead of a horde of ants, with an average weight of 0.0001 ounces (0.003 grams).

Whether due to their diversity, their beauty, or for other reasons entirely, Darwin was as big a fan of beetles as most of us are of, well, The Beatles. Despite being told by his father that his preoccupation with the natural world kept him from focusing on anything of importance, and that he would end by being a disgrace to his family, Darwin became an influential scientist during his lifetime and the most famous biologist of all time.

And it all started with his love of beetles.

Anyone with an interest in the diversity of life's forms should take an active interest in beetles.

Courtship Feeding Rituals

February 16, 2009

There are those who refer to Valentine's Day as "Single Awareness Day," but I like this holiday and think that it has something to offer everyone. Depending on our perspective, we've all been charmed or nauseated by couples exchanging flowers, candy, and sweet nothings, and sometimes even feeding one another this past weekend. Watching lovebirds place morsels of food in one another's mouths can cause some people to roll their eyes or gag, but as a biologist, I find this behavior fascinating.

I am always delighted to see this feeding ritual performed with wedding cake by newlyweds, as long as it is kindly done. It offers a tangible reminder that people are a part of the natural world, because feeding a partner is not limited to humans but rather occurs in many species, particularly birds. The term biologists use for this behavior is "courtship feeding."

Courtship feeding occurs in many birds, including gulls, hawks, cuckoos, owls, terns, falcons, cardinals, kingfishers, and quail. It can happen when pairs are just becoming established or when they are already a couple.

In some species, the more food a male gives a female, the more eggs she lays, and the heavier those eggs are. Females can't always fly or hunt right before they lay their eggs, so the males' provisioning is critical to reproductive success. Besides the nutritional benefits of courtship feeding, it may strengthen the pair bond, reduce aggression between the male and female, and may be an inducement to mating.

Gallinaceous birds (e.g. pheasants, bobwhites, jungle fowl) even have a call they use to get a potential mate to approach them when they have food to offer. They perform this call in combination with a display. The call and display together are called "tidbitting" and are almost always successful at getting the object of their affection to approach and take the food.

The transfer of food varies across species. Cardinals deliver seeds directly to their mates' bills while birds of prey present an item such as a lemming, mouse, bird, or snake to a female, who takes it from the male's talons or bill. In the Northern Harrier, the food is given and received in flight with the male either transferring it to the female's talons from his own, or, if the female is flying below, by dropping the item for her to catch, a maneuver that requires her to turn over. In gulls, the male vomits his gift of partially digested fish and squid, depositing it on the ground in front of her. Fine food, attractively presented.

The next time you see courtship feeding in our species or any other, consider what it means to you. Do you take a moment to appreciate observing food being shared in a loving way? Or, do you get an unstoppable urge to comment on how animal-like it is, or even to mention that some pairs feed each other with food that has been regurgitated? With this full spectrum of possibilities available, I stand by my statement that Valentine's Day has something to offer everyone.

There were objections to my inclusion of the feeding of wedding cake by newlyweds, but as an animal species, humans share many traits and behaviors with other species, including feeding our mates.

Fish Can Change Sex
March 2, 2009

"The birds and the bees" means so much more than a discussion of birds and of bees. In fact, a properly thorough discussion of reproduction must include fish, because fish reproductive behavior definitely involves some of the most bizarre forms.

In many fish species, an individual can be both male and female. Salmon, for example, contain both sperm and eggs simultaneously, allowing them to spawn with any other salmon that they happen across. This is an important capability in a dark, watery environment in which encounters with other members of the same species can be rare. Individuals that have both male and female reproductive parts at the same time are called simultaneous hermaphrodites. The term "hermaphrodite" comes from the minor Greek god Hermaphroditus, who was the son of Hermes, the patron of boundaries and travelers who cross them, and of Aphrodite, the goddess of love and beauty.

More common than simultaneously hermaphroditic fish are the sequential hermaphrodites. These are individuals who are one sex as a young fish and then change into the other sex later in life. Not every member of sequentially hermaphroditic fish species will undergo a sex change, but every individual has the capability of doing so. Most sequential hermaphrodites are marine reef fishes such as wrasses, parrotfish, angelfish, moray eels, damselfish, and some gobies. In most species, a fish can only change once, but a rare few can reverse the change, too. Changing from male to female is called protandry and changing from female to male is called protogyny. Protogyny is more common than protandry.

Changing sex allows for better reproductive success, measured as the number of offspring an individual can produce. The social structure of the fish determines the particulars of each species' curious life history. In protogynous fish species (those that change from a female into a male), the adults form harems with a single male accompanied by many females. The largest male in the area mates with all of the females, fertilizing all the eggs produced by the members of his harem. Smaller males are out of luck, reproductively speaking, and unlikely to have much reproductive success. On the other hand, females, no matter how small, are able to produce eggs and mate with the male, thereby achieving some reproductive success. Although all females get to mate, no single female can have as many offspring as the male of the group can. When the large male dies, the largest female in the harem changes into a male, which results in her having more offspring than would have been possible if she had remained a female.

In protandrous species, the social system is most often the reverse—one large female living with multiple smaller males. If the female dies, the largest male in the group will undergo a sex change and become female, which improves his reproductive success relative to remaining a male. Sex changes from male to female or from female to male are commonly reported to take 4 or 5 days.

Truth may be stranger than fiction, but nothing is stranger than fishin'.

This article came out during the letter *S* week at my son's preschool, and he offered the word "sex" as an example of an *S*-word—not surprising as we had discussed the article at breakfast. The teacher just said, "Yes, that word starts with *S*."

People Can Mimic Bird Vocalizations

March 16, 2009

I knew my husband was an excellent vocal mimic of birds early in our relationship. He had returned from a bird survey quite chagrined that his version of a bird song had been taken for the real thing by the group leader.

In an attempt to call in more Black-capped Chickadees to the area, he had whistled the two-note "phoe-bee" song that is commonly heard in the north woods, especially in the spring, and is made mainly by males to defend their territories and attract females.

With no intention of fooling anybody except the chickadees themselves, he was alarmed to see the graduate student in charge of the outing record his whistle as an actual bird song. The kicker was that this graduate student was in the final stages of writing up her PhD dissertation, the subject of which was the two-note, clear, whistle-like "phoe-bee" song of the Black-capped Chickadee. Clearly, it is possible to make animal noises that are so realistic that even an expert can be fooled. He doesn't know if the birds were also taken in.

On another occasion, we had just driven 1,000 miles from Wisconsin to his parents' summer place in the mountains of North Carolina, and as we got out of the car, we heard the call of an Eastern Screech Owl. This call is a long, descending, and slightly eerie whinny of a call used most often by males during courtship. Without even closing the car doors, much less unloading our gear, we began to move in the direction of the sound with my husband answering the call.

Even after hearing about the chickadee debacle, I was stunned by the accuracy of the call. Each time I heard the call in the distance and my husband's answering call, I was compelled to comment. "That's amazing!" "You sound exactly like that bird!" "I literally can't tell the difference!" "Wow, that's unreal!" (Besides the fact that his call was such a perfect imitation of the one we kept hearing from the unseen bird, you have to realize that we had been dating for less than 5 months, so I was especially prone to be impressed.)

After about 20 minutes, it was obvious the owl he was communicating with was coming closer. The calls between the two were occurring more frequently, and the answering calls were growing louder. I was so excited to get a close-up look at an Eastern Screech Owl, and ready to be awed by my then-boyfriend's pied-piper-esque abilities. Finally, after an additional 10 minutes of careful calling and answering, the two callers came in sight of each other. We were face to face, not with a majestic bird of prey, but with my husband's identical twin brother (who amusingly enough was studying the Eastern Screech Owl for his master's thesis). Did I mention that the calls were indistinguishable from one another? Instead of being awed, I found myself in hysterics—either way, a positive way to feel when spending time in the woods with the man in your life, and his brother.

My husband continues to whistle and call from time to time while we are out birding, but I avoid being fooled by regularly asking, "Was that you or a real bird?"

The Trouble with Bugs

March 23, 2009

As we watched people taking their places at our wedding, we realized that a non-traditional way of balancing the seating arrangement would have been "biology geeks" on one side of the aisle and "computer geeks" on the other. Though this distinction covered our entire wedding party and a sizable portion of the guests, there was too much common ground for any conflicts to arise. For example, we could all agree that the term "bug" is guaranteed to be associated with problems.

For me and for my fellow biologists, the misuse of the word "bug" is a predictable source of irritation. Technically, a bug is an insect in the suborder Heteroptera, the so-called "true bugs" such as stink bugs, water striders, and bed bugs. This is at great odds with the erroneous but common usage of this word to refer to any creepy crawly creature. You'd be hard pressed to find anyone who has taken even a single entomology class who can hear anyone refer to a spider, fly, tick, or any other non-Heteropteran as a bug without experiencing at least a mild cringe.

For the computer crowd, the link between "bug" and a sense of the problematic is even more direct since the term is synonymous with a glitch of some kind in either hardware or software. A famous story involves an insect that was found in a malfunctioning computer. One day in 1947 at Harvard University, the primitive computer called the Mark II Aiken Relay Calculator was having problems. Investigations into the difficulties led to the discovery of a moth trapped in one of the relays. This story is sometimes cited as the origin of the use of the terms "bug" and "debug" in computing, as well as more broadly in engineering and technology.

I love this well-known story, which is why I was saddened to learn the truth. While an insect was found in a relay on the computer, it is not true that this is where the term originates. In fact, "bug" had been used in this way for many decades. Thomas Edison even remarked in an 1878 letter about the process of fine-tuning his inventions for commercial success that there is always a phase in which "'Bugs'—as such little faults and difficulties are called—show themselves."

The fact that the term "bug" already referred to malfunctions is what made the entry in the logbook for the Mark II on that particular day so funny. The operators who discovered it, blessed with a sense of irony, removed the moth from the relay, taped it with care into the logbook and wrote, "First actual case of bug being found."

Had there been a trained biologist present to lend diversity to this group, surely there would have been an objection to referring to this moth as a bug. Since we do not worry about how to "de-moth" our computers, I am grateful that a minor little inaccuracy was not allowed to get in the way of a clever joke.

My father, a retired computer science professor, edits all of my columns before submission, and claims he has received quite a biological education over the more than 14 years of doing so. For this column, I turned to him for a lesson involving the computer science information I needed.

Support Your Local Monster

April 6, 2009

Although there are poisonous animals and venomous creatures galore that many people consider monsters, few are officially called monsters. In that rare category is a species that we have right here in Arizona, although not in Flagstaff: the Gila monster.

Arizona is even responsible for part of the species name, which comes from the Gila River basin where the species was first discovered. (Let's assume the "monster" part refers to some other state in which this species lives. We don't want to be greedy when taking credit, since Gila monsters also live in Utah, Nevada, California, and New Mexico, as well as in the Mexican states of Sonora and Sinaloa.)

Actually, the Gila monster is generally docile, and it takes a real effort to be bitten by one. Many folks do make that effort, perhaps in a quest to sacrifice their bodies to science, or perhaps to confirm that human nature leads to many actions that fall into the category of "not a good idea." People who are bitten by Gila monsters often suffer this fate because they tried to pick one up, perhaps enticed by their beautiful coloring—a trait common among poisonous animals.

I recommend against picking up a Gila monster both because of the painfulness of bites and because it is against the law to handle, collect, or kill a Gila monster in the United States and Mexico. High demand for exotic animals as pets and destruction of habitat have made them rare in the wild, which led to them becoming the first venomous animal to be afforded legal protection.

Being attacked by a Gila monster is painful for several reasons. They have claws that mean business, strong jaws, and sharp teeth. The bacteria in their mouths often lead to infection, and venom is injected when they bite. Since eggs make up the majority of their diet, most scientists think that their venom is primarily used to defend themselves, not to subdue or kill prey. Gila monsters are the only venomous lizard native to the United States, and they are also the largest native lizard in the United States, growing up to 2 feet long and weighing as much as 5 pounds.

A substantial part of that weight is in the Gila monster's tail, which stores fat and can be quite large. In fact, members of this species can live off the fat in the tail for several months, and perhaps even years, according to some anecdotes. A Gila monster eats only 5 to 10 times a year in the wild, and its main food is bird and reptile eggs. It will also occasionally eat small mammals, birds, lizards, frogs, insects, and carrion. They sometimes climb trees and cacti to search for eggs.

The Gila monster is such an unusual animal that having the word "monster" in its common name gives the proper impression of this animal as a creature to be reckoned with, and since it certainly exists, we Arizonans deserve to be even more proud of our "monster" than many Scots are of their Loch Ness Monster.

I have yet to see a Gila monster in the wild despite living in Arizona for 18 years. For the record, I spent a month in Scotland and never saw the Loch Ness Monster, either.

The Miracles of *Earth*

May 4, 2009

Seeing wild animals is always a special event, and last weekend, I saw things I never thought I would see, like polar bear cubs emerging from their den for the first time, lions attacking migrating elephants, swordfish swimming in the open ocean, and a humpback whale mother pushing her daughter up to the surface to breathe. That's right, I went to see the movie *Earth*.

Normally, I only get excited about what I see live and in the flesh, but this movie is an exception. The footage was absolutely unreal and raises many questions. I mean seriously, how DO you film a polar bear when you are a mile away from it, and it is swimming for days and then attacking walruses? Do standard film school classes cover the techniques for capturing images of huge flocks of cranes on both their first and second attempts to fly over the Himalayas? Can time-lapse photography really be used to capture entire seasons of change in habitats all over the world? How can the world's economy conceivably be in turmoil when projects such as this are being financed?

I guess with the help of the latest photographic technology, planes, helicopters, a hard-to-steer hot air balloon, and a staff more populous than many towns, the improbable events in nature can be captured for the world to see. Successful hunts by wolves may be horrible for their caribou prey, but seen in slow motion, they appear graceful and captivating.

My favorite part of the movie was watching a brood of new ducks fly for the first time—an event that narrator James Earl Jones described as falling with style. In slow motion, it was charming to watch the ducks bounce multiple times as they landed in the leaf litter below their nest up high in a tree, and adding to the charm were the childish giggles of the kids (and some adults whose names will go unmentioned here) in the theater. Yes, many species of ducks do nest in trees, believe it or not, and not surprisingly, they look a bit silly when they head to the ground sans practice.

Another thoroughly enjoyable scene, shot from above, showed massive numbers of dolphins swimming and leaping in the ocean as they simultaneously headed for points unknown. And, who could resist the penguins waddling and sliding about as though their purpose in life was to be ridiculous in order to amuse? (People could say the same about me when they watch me dance, but they are usually too polite to do so.)

The music for this film is appropriately dramatic, which enhanced the majesty of the images. The result is that unless I imagine the score playing in my head, all of my animal sightings since seeing *Earth* seem oddly mundane. Of course, it's hard for a robin in my yard, a squirrel running at the park, a grasshopper on a flower, or an ant crossing the sidewalk to compare with a great white shark swallowing a sea lion.

Go see the film. And that suggestion has nothing to do with whether or not I own stock in Disney.

Such extraordinary footage of animals in the wild is only possible with modern technology as well as nearly endless supplies of time, patience, and money.

Puffins—A Favorite for All Ages
May 25, 2009

If you run across a preschool child who takes a break from doing cartwheels to reply "a puffin scientist!" when asked what he wants to be when he grows up, then you've either met my youngest son or his soul mate. Most 3- and 4-year-olds are into dinosaurs, horses, lions, or dogs, but he has inexplicably chosen the puffin as his favorite animal.

Puffins are definitely beautiful birds, and a common favorite among adult bird lovers. These seabirds have such colorful bills that their nickname is "sea parrot." Most people think that puffins resemble penguins because they are black and white and waddle a bit clumsily on land. However, puffins and penguins are actually not closely related. All puffins are members of the order Charadriiformes and in the Auk family, whose scientific name is Alcidae. There are only three species: the Horned Puffin, the Atlantic Puffin, and the Tufted Puffin. In contrast to penguins (order Sphenisciformes, family Spheniscidae), which live in the southern hemisphere, puffins live exclusively to the north of the equator.

Puffins spend most of their life at sea. To get good sightings of them usually requires being out on a boat. They do come on land to nest, but their nesting sites tend to be on islands where the risk of predation from animals such as weasels and foxes is lower.

One of the characteristics of a good nesting site is being on a high cliff. The steepness of the slope is critical because puffins are not the best fliers, and they can more easily launch themselves into the air from a high cliff than from anywhere else. Jumping off a cliff to gain speed is one strategy for getting started in flight. To take flight from the water, they perform a long running start to gain enough speed, sort of like the way big jets need a long start on a runway to gain enough speed to become airborne.

Most birds have long wings and hollow bones, traits that help them fly. Puffins have almost solid bones and stubby wings, traits that help them dive 50 to 200 feet underwater to catch the fish and zooplankton that make up their seafood diet, but prevent them from being agile fliers. The lighter bones and longer wings of better fliers would prohibit puffins from diving effectively.

It is amusing to watch puffins face the challenges of landing. They usually fly above the place they want to land and then stall in mid-air. They spread their wings wide, stop flapping them, and if they have judged everything accurately, fall straight down onto the ledge of their nesting site. It's not unusual for a puffin to turn away at the last moment and circle around for another go at landing. Landings are often messy, tumbling events, and many times puffins look disheveled afterwards, requiring careful preening to put their feathers back in order.

Being a puffin lover myself, I am perfectly happy with my son's great interest in this type of bird. All I want to know at this point is whether any schools that train you to become a puffin scientist also offer gymnastics scholarships.

All members of our family continue to be big fans of the puffin!

Cows, Ants, Termites, and Me
June 22, 2009

While I was living in Wisconsin and expecting my first son, I felt a kinship with the cows I drove by on the way to work. We seemed to become more similar in size each day. Rather than being alarmed by this realization, I embraced my new shape. It helped to be secretly grateful that my pregnancy did not cause my form to change as drastically as some insects' bodies do.

In many areas of the world, ants have evolved ways to survive the dry season when food is scarce. The honey pot ant makes use of an unusual division of labor only possible in a large social group. They feed some of their workers huge quantities of water and nectar, causing their abdomens to swell so much that they become unable to move around. They actually lose their mobility as a result of getting too heavy. They just hang in their nest like living pantries storing food for the colony. When the dry, flowerless season arrives, the rest of the colony uses them as living food sources, sort of like ant silos. In

Australia, the indigenous people knew about honey pot ants and raided the colonies for a special sweet treat.

There are insects that get so enormous when reproducing that the idea of pregnant women's growth seems absolutely insignificant in comparison, and even the body shape of the honey pot ant ceases to amaze. In an extreme form of body distension, queens of some social insect species expand enormously during the reproductive phase of their lives. Army ant queens can have abdomens so swollen with eggs that they can lay up to 300,000 eggs in 2 weeks, or over 21,000 daily. Some termite species have queens that are so specialized for laying massive quantities of eggs that they can do little else. These termite queens take from 2 to 12 years to swell up for maximum productivity and their abdomens reach unimaginable sizes. The term for this excessive abdominal swelling is physogastry. Termite queens can lay 40,000 eggs a day, but are unable to take care of themselves, relying on the workers in their colony to feed and care for them.

For comparison, if the average 125-pound woman gains 25 pounds when she is pregnant, that is a 20 percent increase in weight. The average physogastric termite grows to 500 times the size of a new young queen. That's not a typo. She really gets to be 500 times bigger! (By the way, if any woman in your life is pregnant, it's best to assume that whatever size increase she is experiencing is "average." Never commit to any comment that implies otherwise.)

Though my pregnancies are long past, seeing a picture of honey pot ants still reminds me of feeling as though I was the size of Texas. Looking back objectively, I'm pretty confident that I never got any larger than Connecticut. And while I can happily observe cows with fond memories of feeling a bit like one, thoughts of termites are always, and will always be, unwelcome.

The other women in my birthing class thought it was so sweet that my husband called me "Honey Pot" while their husbands called them cows, but none of them knew the real story behind my nickname.

Sweat as a Defense Mechanism
July 6, 2009

Glowing, perspiring, sweating: whatever you call it, there's an awful lot of it happening now that summer weather has finally arrived in Arizona's mountains. Generally, sweating is considered undesirable, even though we all know that its purpose is to cool us down and prevent overheating. Besides making you feel absolutely miserable, overheating can cause serious trouble, including permanent brain, kidney, and other organ damage, and, in extreme cases, death.

In addition to the cooling advantage of sweating, a paper in the scientific journal *Biotropica* reveals that sweat can deter wasp stings. The paper, written by Allen M. Young, is called "A Human Sweat-Mediated Defense against Multiple Attacks by the Wasp *Polybia diguetana* in Northeastern Costa Rica." In it, the author discusses the ability of human sweat to suppress stinging by wasps.

The paper describes the author's observations of a man named Ramon Morales handling an active wasp nest without inciting them to sting. Morales rubbed his left hand into his right armpit to coat his hand with sweat. He then slowly cupped his sweaty hand around the wasp nest, removed it from the banana leaf to which it had been attached and held the nest in his hand for a few minutes.

Most of the wasps crawled over the nest and his hand without attempting to sting him. This is highly unusual because in situations that threaten their nests, they are generally quick to sting. The author describes the wasps as appearing "drugged." None flew to the nearby observers or to any exposed parts of Morales' body. After holding the nest for a few minutes, he broke it open, revealing the comb and young inside. Destroying the nest failed to elicit any signs of attack, which is, again, highly uncharacteristic of these wasps. After the 8 minutes of destroying the nest by pulling it apart, Morales put the nest on the ground, and all the people walked away. At this point the wasps got very active.

These wasps are closely related to the wasp I studied for my doctoral dissertation, and since I was studying their defensive behavior, I know how easy it is to get them to sting and also how much it hurts when they do. Therefore, I am impressed not only by how effectively the sweat prevented stings, but also how confident

Morales had to be in order to perform this demonstration with a nest containing more than a hundred wasps.

Other people in the area claim to have the same ability, suggesting that some people are able to suppress attacks by wasps. The fact that the wasps seem sluggish suggests that whatever component of human sweat is able to deter wasp attacks does so not by being repulsive but rather by actively suppressing the attack behavior that includes stinging.

Little is known about how birds and other vertebrate predators that break open wasp nests to eat the brood protect themselves from getting stung or even how many stings they receive. Being stung by wasps is such a horribly painful experience that although feeling hot and sticky may be unpleasant, it's intriguing to consider that the sweat making you feel that way may have powerful and beneficial properties.

Who knew that human sweat was such a powerful substance? The chemistry behind its power to prevent wasps from stinging remains unknown.

Cats Like to Keep Their Dignity
July 20, 2009

I have worked with dogs for over a dozen years, and understanding them gives me great joy. I love knowing that our two species have associated for thousands of years and even evolved to communicate effectively with each other. There's no denying that dogs and humans share a strong social relationship.

Cats on the other hand, are another story. I love their mystery, and the fact that so much remains unknown about them. You can hardly turn on any nature channel on television without seeing amazing footage of wolves or dogs. On the other hand, when was the last time you happened to flip through the channels and find a show on house cats or their wild progenitor, a Middle Eastern wildcat? We know so little about cats, despite the millions of pet cats out there.

One thing we do know is that their dignity is part of their charm. Yet, when their dignity eludes them, they are even more endearing. I had a cat years ago who once landed on a table as she had many times before, only to slide right off unexpectedly and at high velocity along with the cloth napkin she had landed on. As she flew through the air, her whole body was disorganized, with all four legs sticking out in different directions. Her tail, body, and whiskers were bent in odd ways. She landed neatly, as cats miraculously do, and commenced immediately to lick her paws, stretching them out elegantly one at a time. Grooming is a common displacement activity in all animals, and is often done in times of stress. My cat's chaotic flight seemed to have unsettled her enough for her to groom, yet her facial expression and body posture remained serene, as though nothing unusual had occurred.

Another cat I knew found himself in an even more ridiculous situation. Nipper belonged to relatives in

Vermont who live within sight of the Connecticut River. My cousin Sarah was canoeing on the river when she saw an unidentified white blob on a floating log. Approaching closer, she saw that it was a cat, and thought to herself, "What a stupid cat! That's someone's dumb cat stuck in the middle of the Connecticut River!" Getting closer, she realized that it was her own cat, who for reasons that will forever remain a mystery, was drifting helplessly far from shore. I like to imagine Nipper seeing Sarah and thinking, "It's about time you showed up!" with that feline attitude of expecting us to do whatever they desire at precisely the moment that works best for them.

Nipper meowed anxiously and Sarah paddled right up next to the log so he could get in. Regrettably, he seems to have misjudged how high he needed to jump to clear the gunwales and missed, landing with a splash in the water. Sarah scooped him up immediately and hauled him into the boat. She still considers this image of him standing uncomfortably in the bow of the boat to be her favorite memory of him. Wet and bedraggled, Nipper looked absurd, angry, and certainly undignified, yet still adorable.

I have many fond memories of visiting Nipper when I lived in New England, many of which revolve around watching my dog Bugsy try to play with him. Bugsy was beneath Nipper's dignity and they never became best buddies.

Animals Do the Nastiest Things

August 3, 2009

One great advantage of knowing about animal behavior is that the intricate knowledge of all of the gross things that animals do makes it easy to impress most children. With heartfelt apologies to anyone reading this as they enjoy breakfast, here are some of the most stomach-turning behaviors known.

Dogs are among the species who lick their newborn offspring to stimulate defecation. Of course, these same adult dogs have recently consumed placenta immediately following the birth of the puppies. Though dogs might be our best friends, there does seem to be some component of opposites attracting or at least diversity being valued within our relationship.

Regurgitation as a form of feeding is widespread in the animal kingdom. A lot of social insects such as bees, ants, and wasps do this. Wolves are also well known for feeding their puppies already-digested food. Perhaps the most famous regurgitating animal is the gull, which may be because of actor Jim Carrey's inelegant portrayal of this behavior in *Ace Ventura: Pet Detective.*

Of course, letting what was inside the body go to the outside is not limited to parents feeding their offspring. Many flies eject digestive enzymes onto their food and swallow it after it is at least partially digested. It's not clear how different Emily Post's books on etiquette might have been if humans did this. I suspect she would have spent less time explaining which fork to use and devoted some of those pages to discussing the proper time intervals between spitting on your food and consuming it so as not to bother anyone with the sight of over-digested food. And sea stars actually eject their entire stomach onto their food to digest it and then pull the organ back in once dinner is ready. It's hard to decide which is more disgusting—regurgitating the contents of your stomach or regurgitating your actual stomach.

There are a great many insects that suck blood, most notably the mosquito and the flea, though horse flies, deer flies, and black flies enjoy the same cuisine. Of course, the vampire bat is a blood-sucking mammal that takes more than just the droplet most insects consume, which may be why this behavior is so reviled.

Until recently, blood sucking was considered the most disturbing of the gross behaviors. Now, due to the interest in Bella Swan and her vampire friends from the *Twilight* book series, the idea of blood-sucking creatures is enjoying unprecedented social acceptance. If only the author would write about honorable creatures who are coprophagic, which is a fancy way of saying that they eat poop. This would elevate guinea pigs, capybaras, and rabbits to a whole new level of respect. The popularity that our dogs would enjoy with that sort of press would be astounding considering how much we already adore them despite their well-known participation in this unpalatable habit.

Piles of poop are used by dung beetles to attract mates and as a feeding ground—an odorous singles bar of sorts. Vultures defecate on their own legs to cool themselves off. Apparently poop is a valuable substance throughout much of the animal kingdom. And yet, I'm not pleased when a bird flies overhead and shares some of theirs with me. I guess I'm just weird that way.

I love the beauty of nature as much as anyone, but I find its disgusting side intriguing, too.

Guinea Pigs: The Next Big Thing?
August 24, 2009

Movies and television influence the pets that people get. From *101 Dalmatians* and *Beverly Hills Chihuahua* to *Lassie* and the recently deceased Taco Bell spokesdog, what we see in the media now is what we often see in people's homes a few months down the road.

Disney's new movie, *G-Force,* features highly skilled, specially trained guinea pig spies rescuing the world. The likely result of this action-packed movie is children begging their parents for pet guinea pigs, which could in turn lead to spur-of-the-moment adoptions of guinea pigs. Impulse buys are never a great way to acquire a pet of any kind.

In a press release, the American Humane Association warns against spontaneous purchases of guinea pigs and cautions that although they can make great pets, "guinea pigs are not good 'starter pets' for young children. They are very fragile." American Humane recommends doing research before adopting a guinea pig, or any other animal, as a pet.

So, if you feel that a guinea pig is an animal you might like to have as a pet, what do you need to know about them before making your final decision?

To start with the basics, guinea pigs live about 5 to 8 years and they reach a length of 8 to 10 inches. They are active during the day and typically sleep about 5 hours per night. Because they require gentle handling to avoid injury, they make better pets for children who are at least 8 years old than for those who are younger. Lifting them requires the use of two hands and they need to be gently supported both at the shoulders and under the back end.

Keeping a guinea pig costs about $40 per month. They require bedding, food including a daily dose of fresh food and vegetables with regular additions of hay, vitamin supplements, and water. Because guinea pigs' front teeth grow throughout their life, they need lots of treat sticks and chews to keep their teeth from becoming overgrown.

Cages to house guinea pigs must be at least a foot tall, 2 feet long and a foot wide. Their homes must be cleaned regularly, which means at least once a week, and preferably more frequently. Their nails need to be trimmed regularly, and they need to be seen by an exotic animal veterinarian annually, and for additional visits if they show any signs of ill health such as bald patches (except for hairless areas behind the ears, which are normal), lethargy, ceasing to eat and drink, losing weight, overgrown front teeth, sneezing, or diarrhea.

This species is social, and as a result, they do best if they live with a guinea pig companion. They tend to be very affectionate, fond of attention, and quite playful. They need social interaction and lots of attention every single day.

Guinea pigs can make great pets, as long as you know what to expect from them and what they need from you. Even a small pet requires a considerable commitment so they can have the love, attention, and care they deserve. Should you see Disney's movie and decide after doing your research that guinea pigs are the right pet for you, may the force be with you.

My internet went down while I was researching some basic guinea pig facts for this column, so I went old school and borrowed one of my kindergartener son's animal facts books. He was looking for it while I was working and was not pleased to learn that his own mom was the thief who had caused his distress about the missing book.

It's a Bird, It's a Plane, It's a What?
August 31, 2009

Our son, Brian, was not yet 2½, which may explain why the firefighter did not think it likely that the little guy had, in fact, correctly identified the bird. The firefighter was only trying to keep Brian calm after our car accident, which is why he pointed up at the sky, and said, "Look, there's a buzzard up there." Brian followed his fingers, took a good look at the bird himself, laughed, and announced, "That's a Turkey Vulture!" He most likely identified it by the characteristic way it held its wings up in a V shape while rocking back and forth in flight.

Bird identification mistakes happen to everyone. One field guide has a page called "Confusing Fall Warblers" and if that doesn't cause a dip in your confidence, you're just not paying attention. Less troublesome birds can be confusing, too. For example, you often hear the following type of comment: "It's a crow, no raven, wait, yes, a crow, just as I thought." Crows and ravens are actually fairly easy to tell apart in flight by the shape of their tails. The end of a raven's tail is rounded while the end of a crow's tails is straight. Though the shapes are distinctive, it's not always easy to get a good view of the tail. The Common Raven found here in Flagstaff is bigger than our only crow, the American Crow, a fact which would be considerably more helpful if the birds would cooperate by standing near a ruler or yardstick, which neither species is prone to do.

Among serious birders, perhaps the most difficult identifications are of flycatchers in the genus *Empidonax*. Even in a field guide, many of the 10 or so species in North America look almost identical, and they are even more baffling in the field.

Theoretically, subtle differences in tail length, size and bill shape can be useful clues. The operative word here is "subtle" and to be honest, I have never confidently identified one of these birds based on those tiny variations. The best, and in my case only, way to positively identify an *Empidonax* flycatcher is by their calls and habitat.

Other errors are easy enough to avoid, but even people who should know better sometimes misidentify birds. I once read this gem of an acknowledgment at the end of a scientific paper on foraging behavior in shorebirds: "Finally, we thank the gracious gentleman with the expert birding skills who kindly pointed out to us that the Semipalmated Plovers we were studying were, in fact, Killdeer." Both of these species are plovers in the family Charadriidae, but Semipalmated Plovers have a single black neck band, a small beak, and are about 7 inches tall. In contrast, Killdeer have two black neck bands, are over 10 inches tall, and have a much longer beak than that of the Semipalmated Plover. Whoops. I wonder whether the authors' credibility suffered in the eyes of other scientists after such a gaffe.

For the record, our son was very forgiving of the firefighter's birding error, commenting afterwards, "Well, he knew it was a bird and not a plane."

Brian is now, at age 19, an excellent birder by any measure—not just for his age. For 16 years, I have had far greater confidence in his bird IDs than my own.

Holy Swimming Bat Rays!
September 21, 2009

"Be amazed! You don't see 5-foot-long bat rays every day!" my fellow instructor shouted jubilantly to the junior high school students who were snorkeling with him. Once he told them to be really excited, they changed from a group of people with an attitude of, "Go ahead. Try to impress me," to enthusiastic naturalists reporting on a great find. At dinner that night, they told their friends, "We saw something amazing! Five-foot bat rays! It was awesome," and, "It was so cool. These bat rays were huge and our instructor said that's really rare."

We were environmental education instructors at the Catalina Island Marine Institute, and only a few of the visiting students were lucky enough to see these animals. We took kids out snorkeling every day, and while sightings of horn sharks, sea cucumbers, giant snails, lobsters, purple and orange nudibranchs, señorita fish, gobies, rockfish, orange and neon blue garibaldi, and sea urchins were common, seeing bat rays was definitely special.

Perhaps the coolest thing about bat rays is that they can jump out of the water. On occasion, they can be seen leaping into the air or skimming the surface, and underwater they can be in groups of hundreds. The females weigh up to 200 pounds with a wingspan of 6 feet. Males are smaller, with wingspans of up to 2 feet. At the end of the tail, which is long and thin, there is a stinger with venomous barbed spines.

Perhaps to prevent the stinger from injuring the mother during birth (females give birth to live young—from 2 to 10 pups), the pups are born with a sheath around the stinger and the spines are rubbery. Within days, the sheath falls off and the spines harden. The spines remain hard for the rest of their life, which can be more than 20 years long.

Bat rays are black or brown on their dorsal (top) side but white on the ventral (bottom) side. This color pattern is called countershading and it's very common in many animals including dolphins, shorebirds, and fish. Countershaded animals are camouflaged when viewed from above or below. Animals looking up at a bat ray see a white belly blending in with sunlight filtering in from above but those looking down at one see the dark back blending in with the dark depths below. By being hard to see, bat rays are better able to sneak up on their prey, which includes abalone, crabs, lobster, shrimp, clams, oysters, and many species of fish. They are also more likely to elude detection by their own predators such as great white sharks, leopard sharks, and sea lions.

While most of the students were enthusiastic about everything to do with rays and sharks, there was a notable exception. I distinctly remember one student's comments during feeding time for the sharks and rays that we kept in captivity in our fish lab. As we gave them their ration of squid, the student asked, "How come they're eating calamari and all we got for dinner was corn dogs?"

Decades later, I still love to say, "Be amazed!" if I think someone has not fully appreciated how lucky they are to be experiencing a particular wildlife sighting.

How Salmon Find Their Way
October 12, 2009

J. R. R. Tolkien's famous saying, "Not all those who wander are lost," did not apply at an animal conference I attended at Nashville's Opryland Hotel. It was so unbearably challenging to find one's way around this labyrinth that all nomadic behavior could be attributed to people being unable to locate their destination. On the third day of the conference, I came upon a group of people wild with laughter. They were clustered around a map, and every member of the party appeared hopeless or helpless, and one man threw up his hands in a show of complete defeat. Having figured out how to get around some areas of the facility, I offered my assistance. The answer came from the one person who was not laughing too hard to speak: "I don't think you can help. We can't find the session about our area of specialty—orientation and navigation."

Humans regularly fail to find their cars after walking around the mall for an hour, but many animals are adept at traveling with great precision from one place to another, even across incredibly long distances and after much time has passed.

Salmon make an amazing journey in order to reproduce. They are anadromous, which means that they migrate from the ocean to freshwater streams to spawn. They hatch in fresh water, then head out to sea for the majority of their lives and then return to the very same stream of their own origins in order to spawn the next generation. Pacific salmon swim thousands of miles to return to their natal stream. Daily distances traveled can reach 45 miles. Many die on the way, and even the ones that make it can be in pretty bad shape with bruised areas of skin, torn fins, and injured jaws. The females are swollen with massive quantities of eggs while many males appear emaciated due to the fact that salmon cease to feed once they leave the ocean.

It is astounding that salmon are able to return to their stream of origin. During decades of research into this phenomenon, the famous ethologist Arthur Hasler reported that the salmon find the stream they came from using olfactory cues. That is, they find the right stream based on how it smells. He showed that the unique soil and vegetation composition of each stream gives it a characteristic odor that is the same even over many years, that the fish can tell the difference between the odors of different streams, and that the fish can remember the specific odor of their own stream even after years have passed between their migration to the ocean as youngsters and their return to those streams as adults.

The migration of salmon requires extraordinary feats of perception, learning, memory, and endurance. Clearly, humans aren't as proficient at navigation as the average salmon. I am grateful that my species has the assistance of GPS units, maps, compasses, and the ability to ask for directions. All of these techniques work much better for me than trying to sniff my way back to where I was born.

Art Hasler's work on salmon migration and their return to natal streams saved the salmon industry by influencing management programs worldwide. It's a reminder of the importance of basic research, whether the practical benefits are immediately obvious or not.

Applying Animal Behavior to Real Life

November 2, 2009

Being an applied animal behaviorist is not a well-known career. Many kids want to grow up to be firefighters, dentists, actors, and teachers, but few say they want to be applied animal behaviorists. It's too bad, because the world could use more of them. Applied animal behavior consists of using what we know about animal behavior in practical ways, and there's no shortage of need for such skills.

Though most of my applied animal behavior work involves dogs, I spent years studying insects, so I learned many practical applications of knowing about their behavior.

Many insects fly up to escape danger, a useful fact when doing a catch and release with a yellow jacket that has strayed into our home. Catching a yellow jacket is easier if I trap it against a window with a container and then tilt it to attach the lid to the lower side while the wasp is flying up trying to find an escape route. If I were to hold the container upright and try to put the lid on, I would be far more likely to have the wasp fly away before I could take it outside.

Though I love insects in general and many types specifically, deer flies will probably never endear themselves to me. I can't stand their buzzing around my head and I'm even less fond of the unpleasant, though not truly painful, bites they deliver. These insects are attracted to dark colors, and since my hair is dark brown, they often choose to be near me. They tend to fly up to the tallest warm-blooded animal around, so I've been known to get close to another person, lower my head to just below theirs, wait until the fly is interested in the other person's head, and then duck down as I move away, completing a deer fly transfer.

Many other insects are attracted to dark colors, so wearing white or other light colors is one chemical-free way to minimize the number of mosquito bites one receives. Wasps and bees are more likely to sting dark objects than light ones, so light clothing can provide some protection from attacks. This is one reason bee suits are white, along with concerns about overheating in the sun. (The tendency to attack dark objects may be related to these insects' ability to deter predators with just one or a few well-placed stings around the eyes, which are dark in most vertebrate species, including two common predators of wasp brood: birds and monkeys.)

If you're interested in avoiding bee stings (and who among us isn't?), it's a good idea to avoid smelling like a banana when passing near a hive. There is a chemical in the odor of bananas that acts like bees' alarm pheromone, in that it, too, incites bees to attack. So, from now on, stop going near beehives when you are either eating or wearing bananas.

Avoiding bites, stings, and other annoyances caused by insects relies more on knowledge than on equipment—more proof that education is never a waste.

Applying what I have learned in my work with animals is a continuing theme in my life. It's such a big part of the way I approach so many situations that I wrote a book about what I've learned working with dogs that I apply elsewhere called *Treat Everyone Like a Dog: How a Dog Trainer's World View Can Improve Your Life.*

Ants Enslave Other Ants
November 16, 2009

Evidence from nature that "truth is stranger than fiction" includes slave making in ants. This bizarre phenomenon involves some species of ants raiding other ant colonies to acquire workers for their own colony.

Ant raids resemble military operations. Raiding columns are several meters wide and the first raiding ants wait outside until the whole group has arrived. Then the entire raiding force enters the nest simultaneously. The adults from the colony being raided are rarely successful at defending their nest or escaping with their larvae and pupae. Members of the raiding colony grab any young that adults from the raided colony are carrying and kill the adults that resist. After the raid, adult survivors from the raided colony tend to the few remaining young and start producing new brood to replace those that have been taken away. The raiding soldiers head back to their own nest carrying the stolen larvae and pupae that will work for their captors for the rest of their lives.

The great myrmecologist (ant scientist) Pierre Huber first described slave making (dulosis) in 1810, and it has been a popular area of study ever since. Charles Darwin himself pondered the subject at length and proposed the first idea for how this phenomenon came about—a hypothesis since confirmed.

In *On the Origin of Species*, Darwin suggested that the first step in the evolution of dulosis was ants raiding colonies of other species of ants for food. When they brought pupae back to their own colony, some of them were kept in storage chambers, where a few pupae remained long enough that they emerged as adults before they could be used as food. These captured individuals performed work for the colony that had collected them, which was advantageous for that colony, resulting in an increased likelihood of raiding other colonies to collect pupae to serve as future workers rather than as food. There are present-day ants that do collect pupae from other species for food, and occasionally adults of these raided species are found working within the colony that collected them.

Representing a further progression from this initial stage are species whose colonies regularly have slaves working in their colonies but sometimes do not. It's not unusual for multiple species of slaves to be present within a single slave-raiding colony.

The most advanced slave-making ants are those species that are completely incapable of performing many tasks necessary for the maintenance and reproduction of their colony. With no means of producing their own workers, the soldiers must acquire workers by capture. These soldiers have sickle-shaped mandibles that very effectively puncture and kill adult ants in the colony being raided. Though powerful weapons, these mandibles render the ants who possess them incapable of excavating the ground to build a nest, of caring for the larvae and pupae in a non-injurious way or moving them safely around the nest, or even of acquiring food. Slaves perform these tasks.

Humans are lucky that the worst that happens when we are raided by ants is that they make off with parts of our picnic!

The evolution of dulosis is a great example of how behavior can evolve from other behavior. That's important to understand because arguments against evolution often claim, falsely, that there is no way such behavior could have evolved, yet many steps make even the most complex behavior understandable in terms of its origins.

The FreshwaterAlphabet Encyclopedia
December 21, 2009

In Madison, Wisconsin, where I went to graduate school, there's a building on the shore of Lake Mendota containing the lake lab that houses the scientists who study that lake and many others. Though I studied tropical terrestrial systems, years of contact with the "lake people" taught me a great respect for freshwater systems and the many organisms that live in them. That's why I was so enchanted to read a new book for children that came out this year.

It's neither your ordinary encyclopedia nor your ordinary alphabet book. It's Sylvester Allred's *The FreshwaterAlphabet Encyclopedia,* and it's really something special. With a short introduction about the importance of freshwater habitats and the people who study them (limnologists is the technical term, not the informal "lake people" I tend to use), the book gets right to the alphabet and the fun for kids begins.

And kids love it! The information comes from a scientist, which makes all the difference. As a scientist myself, I'm always thrilled to find books to read to my kids with up-to-date, accurate information that captures their attention. Allred's latest book is one such rare find with great gems of information such as these:

Alligators are descended from the dinosaurs. Backswimmers are aquatic insects that actually do swim belly up on the surface of the water. Bald cypress trees are conifers that shed their needles. Dippers are birds that swim underwater. Fossil dragonflies had 2-foot wingspans. A common name for giant water bugs is "toe biters." Killifish eat mosquito larva and are used for mosquito control. Muskellunge grow to be 8 feet long and weigh 70 pounds. Shrikes impale their food (snakes, insects, small mammals, and birds) on thorns or spines and then eat them. Orb snails have one lung and one gill, and are thus able to breathe in air or water. Pitcher plants and Venus flytraps are carnivorous plants that digest the insects that they trap. Sturgeon can live to be 125 years old and have hardly changed in 100 million years, looking very like their ancestors who lived during the time of the dinosaurs. Zetek treefrogs live in pools of water that collect in bromeliad plants living high up in tropical trees.

Just check out the unusual set of animals at any letter (my favorite is "H," which covers Harlequin Duck, Hellbender, Horsetails, and Humpback Chub, but I also like "L," which consists of Lamprey, Loon, Leech, and Lotus Flower.) And as far as favorites, it includes one of the animals I've always liked best: the Jacana is a bird with feet shaped to allow it to walk on top of plants on the surface of the water.

Allred is a local author whose full-time job is as a professor of biology at Northern Arizona University. I met him at a book signing at the bookstore during the university's family weekend earlier this fall. Our copy of *The Freshwater Alphabet Encyclopedia* is autographed personally to my younger son. It says, "To Evan—Freshwater animals and plants are important to learn about. Enjoy." We've learned. We've enjoyed. Now we recommend.

Though my kids are teenagers now, I still have this book in the small pile of keepers from their childhood.

Snow Up to Their Ears

February 1, 2010

Like everyone in Flagstaff, lately I have thought of little else except snow. Okay, maybe I branched out and thought about snow shovels, snowmen, snowshoes, school snow days, snow forts, and snowballs, but that's about it. So, it was natural that when I was thinking about animals, the snowshoe hare came to mind.

I felt envious of snowshoe hares each time I sank up to my hip in the snow, because these animals are so adept at walking across the surface of snow. In fact, they get their name from the large, furry back feet that help them stay on top of snow rather than sinking into it. In addition to being light on their feet in the snow, snowshoe hares are agile and quick. This is one major way that they avoid their predators, which include lynxes, foxes, bobcats, coyotes, weasels, and birds of prey.

Another way that snowshoe hares avoid predators is camouflage. By blending in with their environment, they are difficult to see. Their winter habitat is snowy, and during this season, the hares are mostly white, with black on the tips of their ears. In summer, their fur is brown. The gradual change of color, which takes approximately 10 weeks, means that the hares blend in during the changes of season as well, when the ground is only partially covered in snow.

Snowshoe hares change color in response to the day length, which is sensed in the pineal gland in their brain. Shorter days in the fall trigger a change from brown to white while spring's longer days signal the reverse—a change from white to brown. With climate change, there are fewer days of snow cover in many parts of the snowshoe hare's range. Though the hares change colors at the same time as they have historically, they are now more often colored white while in a habitat that is devoid of snow. As a result, they are more vulnerable to predation in spring and fall than they used to be.

Snowshoe hares are unusual because their color change is a defense mechanism. Most color changes in the animal world benefit animals by helping them to reproduce. For example, the colorful plumage of many male birds helps them attract females to mate with, and the most brightly colored fish are also often the most popular with the opposite sex.

It might seem as though, with the prolific reproduction so typical of hares and rabbits, snowshoe hares would take over the world. Instead, they are famous for the extreme changes in their population. The number of snowshoe hares fluctuates with populations in Canada's Yukon going from a high of 200–300 individuals per square kilometer to a low of 7 per square kilometer. These boom and bust cycles, which may be caused by disease, have a huge impact on the animals that depend on them as a source of food, most notably the lynx.

Now, having devoted some brainpower to snowshoe hares, it's back to thinking about snow. I wonder, could snowshoe hares help me shovel all this snow?

I wrote this article after living through the more than 4.5 *feet* of snow Flagstaff received in less than a week.

Combat Fish Trouble
February 8, 2010

The private lives of many people include the same deep dark secret. Probe into the history of the folks you know and you will discover something shameful. Many of your nicest neighbors, friends, and colleagues are probably fish killers.

Lots of people have the most horrible trouble keeping their pet fish alive. Over the last year, one family I know has mourned the loss of seven fish. They usually have two or three at a time, and sometimes just one dies, but other times, the whole tank succumbs under their care.

There are multiple reasons why people lose their fish. One of the most basic problems is a tank that is too small. Many species require tanks of at least 20 gallons, yet many people's fish tanks are only 5 or 10 gallons, or even as small as 1 gallon. Make sure you get the right size tank for your fish. Or, if you already have the tank, be careful when selecting your fish so that you choose species that will thrive in it.

Water chemistry is often the culprit when fish die. Most important is that the fish be provided either a freshwater environment or a saltwater one, depending on the species. Keeping a saltwater tank is significantly more complicated and work intensive than a freshwater one, so novices are better off with a freshwater setup. Another issue of water chemistry is preparing the tank days ahead of acquiring your fish. Most species require that the tank be filled with water and the filter allowed to run for at least 48 hours before the fish can safely be added. On the issue of filters, many fish have lost their lives because the filter stopped working due to a power outage, or the filter was so dirty that it was accomplishing next to nothing.

Fish require water that is the right temperature. Every year, many people turn their heat down when they go on vacation during the winter, and come home to fish that did not survive the cold. Even small fluctuations in temperature can affect fish, and the smaller the tank, the more vulnerable it is to change.

Perhaps the most common cause of fish death is overfeeding. We have two small guppies in a 5-gallon tank and we feed 1–2 flakes of food each day. Many people, understandingly worried about their fish starving, feed a pinch of food daily, which is as much as 20 times what they need, and in fact well over what they can handle. The extra food leads to levels of ammonia in the water that are too high for the fish to survive.

Of course, along with the causes of death already mentioned, "death due to mysterious causes" may describe the situation that many people face. It is not always possible to determine what caused fish to die. Diseases, competition, poisoning, old age, or stress may be to blame. As so many of us are secretly fish killers, it's no surprise that fish keep some secrets from us.

My husband and kids take care of our fish, and rarely does one succumb.

Made to Finish Your Duet

March 1, 2010

In the movie *Enchanted,* Amy Adams and James Marsden play the fairy tale characters Giselle and Edward, who fall in love at first sight and immediately sing together.

> (Edward): You're the fairest maid I've ever met. You were made
> (Giselle): To finish your duet.
> (Together): And in years to come we'll reminisce
> (Edward): How we came to love
> (Giselle): And grew and grew love
> (Together): Since first we knew love through True Love's Kiss!

The displays of love and harmony seen in our own species have nothing over those in the animal world. Dueting is a phenomenon that appears in many unrelated groups of animals, even those that are not in Disney movies.

Sperm whales duet, and often do so when they are within sight of each other, which means that they are not just singing to locate each other. They even synchronize their calls, responding within seconds to a partner's sounds and copying phrases from them with the same amount of pausing between clicks. The result is a duet with clicks sung together.

Gibbon duets originated from a common song that was then split into male-only and female-only parts, producing a song that requires both a male and a female to be performed in full form.

Rufous-and-white Wrens have songs with more than one function. A pair's duets serve as united calls to other pairs, but if a single female shows up, everything changes. The song becomes less harmonious and more antagonistic. The already paired male will call to attract this other female. His mate will then sing over his song in order to block out his attempts to attract the unpaired female. The male responds to these blocking calls by singing a more complex song that is harder for his mate to disrupt.

Perhaps the most surprising tale of dueting animals is from the world of mosquitoes, where musical harmony comes from adjustments to the rate at which they beat their wings. Mosquitoes duet with their wing beats in such a way as to create specific harmonics from the two sets of wing beats.

Males typically beat their wings at a frequency of 550 to 650 beats per second (550 to 650 Hertz) compared with females, whose wing beat frequency is 350 to 450 Hertz. The rate at which mosquitoes flap their wings is called their fundamental frequency, but in addition to that main buzz, harmonics are generated. Harmonics are multiples of the fundamental frequencies. Mosquitoes can adjust their fundamental frequency so that their harmonics are coordinated. Specifically, they change their wing buzzing speed so that the male's first harmonic and the female's second harmonic are matched, with both being approximately 1200 Hertz.

This is a form of sexual selection in which females are testing the males prior to deciding whether to mate with them. Female mosquitoes are only truly "enchanted" with males who adjust their wing beat frequency to create these harmonious harmonics, and mate most often with the males who act as though they were made to finish the duet.

Since learning mosquitoes duet, I find I dislike them less than I used to.

Blue in the Animal World
April 5, 2010

Skies can be blue and moods can be blue, but to find that color on an animal is as rare as, well, as a blue moon. Many animals are colorful, but red, yellow, orange, and green are all more common than blue when it comes to the rainbow world of showiness so evident across the animal world.

Recently I was lucky enough to see two blue jewels of the animal world. The place that afforded me this rare pleasure was Ometepe Island in Nicaragua. I spent almost two weeks there during the field component of Northern Arizona University's Biology and Forestry course "Tropical Forest Insect Ecology," for which I am one of the instructors.

Tropical excursions allow one to see some of the most unusual and beautiful animals on the planet, many of which visitors have previously seen only on television. For any biophiliac, that is a real treat.

While in Nicaragua, the highlights of what we saw include three species of monkeys (mantled howlers, white-faced capuchins, and spider monkeys), vampire bats, gekkos, iguanas, Great Kiskadees, an agouti, leafcutter ants, a variety of parrots and parakeets, and beetles with an astounding range of shape, size, and color.

With this variety of animals to choose from, my favorites were the two most strikingly blue ones: the morpho butterfly, and the White-throated Magpie Jay, both common enough to be seen nearly every day. The morpho butterfly is an iridescent blue that appears especially bright when these insects flit through sunny patches of the forest. They tend to fly along streams, trails, and roads, frequently flying along the same routes repeatedly throughout the day.

Their flight seems slow and even lazy, but that is deceptive, as these butterflies are extremely difficult to catch in nets. Every entomologist I know of, myself included, has been made to look and feel like a bit of an idiot after an awkward and unsuccessful chase of one. Nobody looks cool stumbling over rocks and undergrowth only to come up with a big whoosh of a miss with the net as the stunning blue of yet another morpho vanishes into the forest.

Just as beautiful, though not as famous, is the White-throated Magpie Jay. This magnificent long-tailed bird can be up to 20 inches tall. It has a blue back and white underside, and the overall impressiveness of its appearance is enhanced by a plume-like crest of black feathers. A black collar across the neck extends up the side where the black forms a border between the bright blue feathers above and the white ones below. Although this bird is lovely to see, its call is as annoying as, and perhaps just a bit louder than, those of other members of the Corvid family, which includes the jays, crows, and ravens. These birds are highly social, even nesting communally, and their squawking and bellowing is usually done in groups as well.

Despite their obnoxious calls, I give this species a blue ribbon for colorful splendor, and seeing them, or a morpho butterfly, is a sure way to shake the blues.

It's unusual for birds, reptiles, insects, or other invertebrates to sport the color blue, but it's especially rare in mammals where the blue whale and the mandrill are the two most famous examples.

Condors at Grand Canyon
April 19, 2010

I have an excellent track record with wildlife sightings. I've been diving with sharks, sea turtles, seals, and sea otters, watched monkeys and Resplendent Quetzals forage in tropical forests, observed giant anteaters and ocelots, seen an endangered Hawaiian monk seal come ashore, helped other scientists capture an anaconda, and even once saw the spout of a blue whale.

Yet, after 5 years of living in Flagstaff with numerous trips to Grand Canyon, I've yet to see a condor. I know, you're probably thinking, "That's weird. I saw one when I went." I know. Everybody seems to have seen them but me.

So, why do I want to see one so badly? Because these birds are wonders of the world. They are magnificent to observe in the wild (or, so I hear) and one reason is their massive size. They can be over 4.5 feet tall, weigh almost 30 pounds and have a wingspan approaching 10 feet. For comparison, the American Robin tops out at under a foot tall and about a fifth of a pound, with a wingspan of 16 inches.

These condors are rare, occurring in parts of California, Arizona, Utah, and Baja California, Mexico. In 1982, condors were perilously close to extinction with only 22 wild birds remaining. A collaborative effort between zoos, government organizations, and private conservation groups resulted in one of the most successful captive breeding programs ever. There are now more than 300 wild condors, though the population is still small compared to the thousands that once lived across the Western states and into Mexico.

Condors are excellent fliers, spending hours in the air riding thermal currents, just as other species of vultures do. They can go as fast as 55 miles per hour and reach an altitude of 15,000 feet. Condors can live over 50 years.

Some people consider the condor a symbol of power. It has been nicknamed the Thunderbird because its great wings were thought to bring thunder to the skies. Other people view them as dirty because they eat carrion. Actually, condors are pretty tidy, cleaning their heads and necks on rocks, grass, and branches after eating and devoting hours to cleaning and drying their feathers.

On our first trip to Grand Canyon soon after moving here from New Hampshire, my husband and our toddler saw their first condor. Where was I? In the car nursing our infant son. By the time they hurried over to tell me to take a look, it was gone. I've been in the bathroom, helping a child with a bloody nose, still driving to get there, or already departed when other people have seen them. Other people count the number of condors they've seen. Me? I can only count the number of times I've heard someone say, "Oh, you just missed it. What a shame. It was so cool!"

Anyone reading this will probably see a condor before I do, since I'm apparently cursed in this regard. I beg you, though, if you happen to spot one when I am there, will you please point it out to me, too? These enormous birds can't all hide from me forever!

When this column appeared, dozens of friends and acquaintances reached out to share that they had seen condors lots of times, and it was so weird that they had eluded me. I have since seen many condors, including some very close sightings. The condor living near me is the California Condor, but I hope someday to see the Andean Condor and fervently hope that no curse will interfere with my efforts.

Ants Rule

May 24, 2010

Ants are one of the most dominant life forms ever. If all the ants in the Brazilian Amazon were weighed, they would be four times heavier than the combined weight of all the mammals, birds, amphibians, and reptiles there. There are an estimated 20,000 species of ants.

Some ants grow fungus gardens to provide food while others forage in massive swarming raids. Some specialize in eating seeds while others subsist on the "honeydew" secreted by aphids that the ants take care of like farmers tend their animals. There are ants that raid other species of ants, enslaving the captured individuals to do their work for them. Other ants excavate nests as deep as 6 meters and there are even species that make arboreal nests out of silk collected from the larvae in their colony.

It's amazing that they are able to accomplish all of this, especially since they have committed such a large amount of their work force to picnic duty, apparently in perpetuity. Another amazing thing is the terminology necessary to explain the lives of creatures so different from us.

Eusocial: Translated from the Greek, this means "truly social." Animals are considered "eusocial" if they meet three criteria: (1) Overlapping generations. (2) Cooperative brood care. Individuals help care for young that are not their own offspring. (3) Reproductive division of labor. Only some of the individuals in the group reproduce while the others work to take care of the offspring.

Castes: These are different morphological types within a single colony with each of the types specializing in the performance of different behavior. In ants, there are three broad castes of females: workers, soldiers, and queens. Females do all the work. The role of males is to mate, and they die soon after doing so.

Trophallaxis: The transfer of foods or other substances from the mouth of one individual to another within a social group. This is very common among ants and other eusocial insects.

Haplodiploid: To understand this unusual genetic system that ants have, along with their bee and wasp relatives, you must first understand the terms "diploid" and "haploid." Humans get one set of genes from Dad and one set from Mom, which means that we have two sets of genes, or that we are "diploid." Haploid organisms only have one set of genes. (The best-known examples of haploid organisms are the gametophyte life stages of ferns and mosses.) So, what makes ants haplodiploid? Ant queens produce males by laying eggs that have not been fertilized by sperm, and they can produce females by laying fertilized eggs. Male ants are haploid, because they only have one set of genes—the one from their mother's egg, while females are diploid because they have a set of genes from their mother and another set from their father. Ants, with haploid males and diploid females, are haplodiploid.

The final two words to know to achieve full ant literacy are "myrmecology," which is the scientific study of ants, and "myrmecophile," which means a species that spends part of its life cycle with ants, but which, directly translated, means "ant lover."

Anyone who learns about ants tends to become amazed by them, and the admiration for their success, social complexity, and diverse ways of making a living is well-deserved.

Who Let the Cat Out of the Box?
June 7, 2010

Today's topic is cat litter boxes. As such, the first order of business is an apology to those of you who are reading this over your morning coffee and an especially heartfelt "sorry" to anyone who, like me, enjoys Grape-Nuts cereal for breakfast.

Cats and people sometimes disagree about how to design the bathroom. This conflict can lead to a cat who uses the floor, the bed, the bathroom towels, or even the Persian rug as a toilet instead of the litter box. The result can be a house that is highly aromatic (and not in a good way). Needless to say, that's not good for the relationship between cats and people.

What works for one cat may not work for others, but the following ideas will encourage most cats to use the litter box from day one and are also some steps to try if your cat is not using it.

Most cats prefer open litter boxes to covered ones. Perhaps the cover makes them feel trapped or it may keep the smell in. They also like clean litter boxes. If you want your cat to be clean in her habits, it helps to model that behavior by scooping the litter box daily and changing the litter often. Other preferences of the typical cat are boxes without liners and those with low sides that they can step over rather than having to jump.

Having multiple boxes in diverse places will make it more likely that your cat will use the litter box. Some cats prefer to urinate and defecate in different places, which requires more than one box. Having a box on every floor of the house helps encourage cats to use them. A guideline that everyone talks about without knowing the origin of the idea is that you should have one more litter box than cats. Many multi-cat households go from elimination chaos to proper litter box use when enough litter boxes are added to meet this requirement.

The type of litter can make the difference between sweet success and smelly failure. Use what your cat is used to, unless she has had a bad experience with it, in which case a change might be beneficial. Most cats prefer soft sandy litter with no added scents, and only 1 to 1½ inches of litter in the box. If she has a paw injury, try pressed paper pellets instead of the sandy kind, which can cause pain on a tender paw.

Litter box location makes a difference to most cats. Put litter boxes somewhere other than where the food and water bowls are. Choose a quiet spot that has low traffic. It should be easy to get to and private, with a way to escape without being ambushed by anyone else in the house.

Some of these techniques may not be convenient, but if you want your cat using the litter box, it's worth paying attention to the kind of "bathroom" she wants. A great resource is *The Fastidious Feline: How to Prevent and Treat Litter Box Problems* written by Patricia B. McConnell, PhD.

Pleas for help with cat litter box issues and inappropriate elimination elsewhere are common for all the feline applied animal behaviorists I know. There are many nuances to this issue, but the general guideline to design a bathroom with the cat's preferences in mind is one important and good approach to solving the problem.

Unseen Jaguars Watch, Wait
July 5, 2010

When I was in Venezuela working on my doctoral dissertation, there were jaguars at my field site, though I never saw any. The jaguar is elusive and often lives in inaccessible habitat, which makes it difficult to find. I regularly saw tracks, and heard the animals roar from time to time, but never laid eyes on one. Now, this may seem depressing, but even the three scientists who were studying this species only saw them a few times a year, so I didn't feel too bad. In fact, it's easy to make the case that it was a good thing I never saw them.

Even compared to other large felines, the jaguar has an exceptionally hard bite. Since humans have such delicate skin, and their muscles and bones are not made of steel, this jaw strength is hardly good news. Their bite is so powerful that they are able to bite through armored reptiles. Jaguars can kill their prey by piercing the skull between the ears and delivering a fatal bite to the brain.

This strength allows them to eat prey not available to less powerful predators. Their prey includes turtles, capybaras, caimans, anacondas, deer, tapirs, peccaries, dogs, foxes, mice, birds, monkeys, frogs, sloths, and domestic animals such as cattle and horses.

Called *el tigre* in many Latin American countries, the jaguar ranges from Argentina and Paraguay to Mexico with a possible population near Tucson. They are the largest feline in the Americas, and the third largest feline in the world after the lion and the tiger. Those living in the northern part of their range are smaller, typically around 70–100 pounds, than those in the south, which can weigh more than 300 pounds. Jaguars can reach 30 inches at the shoulder and achieve a length of 6 feet, plus the 2½ feet of tail.

Jaguars that live in the forest tend to be smaller and darker than those that live in open areas. They do live in a variety of habitats, including the rain forest as well as more open areas and other types of forest. Jaguars are often near water, and like tigers, tend to be excellent swimmers. Both of these species are solitary predators with a stalk-and-ambush style of hunting.

Despite behavioral similarities with the tiger, the jaguar's closest relative is thought to be the leopard. The jaguar is a sturdier, heavier cat with a rounder head, but these two species still resemble each other. Both jaguars and leopards have rosettes on their fur, but those of the jaguar are larger, usually darker, and have a spot in the middle. Roughly 6 percent of jaguars are all-black, though it is still possible to see their spots. These animals are called black panthers, though the scientific term for them is melanistic jaguars.

Even though I never saw a jaguar in all the months I spent in Venezuela, the people studying them assured me that these cats had definitely seen me during my forays into the forests. Isn't that comforting?

While living in Costa Rica in 2013, our family saw a fresh jaguar paw print in the mud and a wet pawprint on a rock in the middle of a stream. We estimate that the jaguar had been there minutes before us and had probably seen us coming . . . and perhaps going.

Rats Make Great Pets

July 19, 2010

Even as a child when snakes and spiders made me nervous, rats were on my list of "good guys."

Later, in college, I once came home to find my roommate standing on our table shrieking. I couldn't understand a word she was screaming, but she seemed afraid of something on the floor, so I jumped up there with her. Once I was finally able to understand that she was terrified of a rat, I got down, feeling silly that I had joined her in the first place. (It turned out to be our neighbor's pet mouse, anyway, and not a rat at all.)

Now as an adult and a behaviorist, I stand by my lifelong love of rats, and often recommend them as pets. Despite many people's fears of them, these animals tend to be quite friendly and people-oriented. They are not generally prone to aggression and can tolerate lots of handling.

Rats make great first pets for kids. They are not nearly as likely to be injured by handling as mice or hamsters, although of course I suggest that children still be supervised and taught how to be gentle with them just as they should be with all animals.

With the endless amount of work in a household with children, it's advantageous that the care of rats is not overly intensive or expensive. Yet, rats are intelligent, curious, and trainable. They can be taught to come when called, to leap from one obstacle to another, to perform back flips, to crawl through tunnels, and to fetch. It's even fairly easy to train them to eliminate in a litter box in their cage because they typically prefer to urinate and defecate in one spot.

Strong bonds can develop between rats and people—often more so than with other small pets such as hamsters, gerbils, and rabbits. In fact, rats typically enjoy lots of interaction with either humans or other rats, which is why many people recommend having more than one rat so that they have company. (Some people caution that having two of the same sex is risky because they might not get along, but getting a pair means they're likely to produce more rats.)

There are several downsides to having rats. They need larger cages than hamsters or mice, although not as big as rabbits or guinea pigs need. A huge drawback of having pet rats is that their life span is short—typically 3 years.

The biggest problem you are likely to have with a pet rat is in the area of public relations. You are bound to have friends who are afraid of them because they consider them dirty, diseased, and disgusting. People will assume you've collected your rat from the nearest sewer (not a good idea) rather than purchased a healthy rat at a pet store, which is the right way to acquire one.

Their negative reputation is undeserved. You'll probably spend time explaining to others the benefits of having a pet rat, since once you have one, you'll be convinced that rats make adorable, lovable pets that are clean, affectionate, and social.

Rats remain one of my top picks for great pets, but they still have a major PR problem and are not as popular as I would like for them to be. And since writing this column, I have learned that acquiring rats through rescue organizations is preferable to obtaining them from pet stores.

Adventures in Pet Sitting

August 2, 2010

My sons recently experienced one of the happiest days of their lives. My husband's graduate student, Ben, brought over his green anole, African green tree frog, and crested gecko so we could care for them while he spends 6 months in Florida doing fieldwork for his PhD. His pets are all "herps," which is a collective term for reptiles and amphibians. (People who study this group of organisms are called "herpetologists.")

So far, the anole has provided the most excitement. Anoles run very fast and can be difficult to chase down and capture. One morning, my sons couldn't find it in its usual spot upside down on the ceiling of the aquarium. They also noticed that the top had been left open, which was my fault. I had opened it to let them put in grasshoppers as food, and when a friend stopped by, I forgot to close it. We looked all over the house, especially any place that the anole could be upside down on a horizontal surface, and in all of our houseplants, but without any luck.

I called a pet store seeking tips for finding our lost pet and was told that they are escape artists, very fast and tend to head anyplace warm, meaning that our anole had likely hit the great outdoors. The kind man at the pet store made me feel a little better when he told me that he has had anoles escape on him countless times. Days later, my husband found the anole on a window, trapped in a spider web. (This validates my lifestyle. If the house were cleaner, that anole might have been lost forever.)

The escape notwithstanding, anoles are fairly easy to keep as pets. They are sometimes known as "starter lizards" because of the ease of their care and their low price. They need a glass or plastic aquarium with a tight-fitting screen on the top. A 10-gallon tank is the minimum size for one or two anoles, with three or more requiring a larger home. Males fight with each other, displaying with the colorful dewlap—fold of skin—under their chin that helps make them such attractive little lizards, so no more than one male should be kept in the group.

They eat live crickets, mealworms, and other insects, and get their water from droplets on leaves, which is why experts recommend that their cage be misted once or twice daily. It is advisable to use a special light to keep their habitat at the proper temperature of 75–85 degrees and to provide the UV light that so many reptiles require for nutrient absorption.

Anoles grow to about 8 inches long, much of which is tail. They are sometimes called the American chameleon, and although they are not true chameleons, they can change colors with a range from brown to bright green.

Absence really did make my heart grow fonder, which is perhaps why since the return of our anole, it is my favorite of the three herps we have despite the considerable charms of both the gecko and the frog.

I'm generally a very responsible pet sitter, and whenever I think of how my mistake could have resulted in losing this anole forever, I feel quite sick.

Tide Pool Diversity
August 16, 2010

One of the coolest habitats for observing the great diversity of life is the rocky intertidal zone, which exists where the ocean meets the land. It's covered with water at high tide, but exposed at low tide when the water recedes. I adore tide pooling, which is going to a rocky beach at low tide and checking out all the life in little pools of water that remain in depressions in the rocks when the tide goes out.

Many of the major phyla of animals can be seen during a single tide pooling excursion. Phyla are groupings of organisms based on their shared evolutionary history and similarities in basic body plan. That is, members of a single phylum are more closely related to each other than members of any other phyla and they share at least a basic level of morphological similarity. Many of the common animal phyla can be found in just about any set of tide pools.

Sponges are in the phylum Porifera, which means "pore bearer." Sponges are radially symmetrical, have no digestive system, and are filter feeders. You can pass an entire sponge through a sieve to separate the cells from each other and they will reform into a complete organism again.

The phylum Coelenterata contains the corals, sea anemones, and jellyfish. These animals have two life stages: the stationary polyp, which can be either solitary or colonial, and a free-swimming medusa. Many of these animals have stinging tentacles. Coral reefs are the ecosystem with the second highest productivity and complexity on earth, behind the tropical rain forest.

The phylum Mollusca contains snails, chitons, limpets, clams, scallops, octopuses, and squids. These animals have bilateral symmetry and secrete a shell. Mollusks are the source of pearls, mother of pearl, and Tyrian purple dye. It is the largest marine phylum, comprising nearly a quarter of all known marine species.

Chordata is the phylum to which we belong. It includes all the vertebrates as well as some animals without backbones such as hagfish, lampreys, tunicates, and lancelets. In tide pools, fish are the chordates likely to be present.

Sea cucumbers, sea urchins, brittle stars, and sea stars are in the phylum Echinodermata. The name of the phylum means "spiny skinned." Their bodies have five basic segments and move with the use of tube feet, which create suction through the use of hydraulics.

Animals in the phylum Arthropoda have an exoskeleton made of chitin and jointed appendages. In tide pools, common members of this group are crabs, barnacles, and shrimp, all of which are in the subphylum Crustacea. Arthropoda includes insects, centipedes, spiders, scorpions, and millipedes and are the most numerous animals on earth.

I love going to the tide pools and seeing animals from these phyla, as well as other phyla, all at once. I'm far from alone in my appreciation of tide pools. Even *Twilight's* Bella Swan had this to say about them: "I loved the tide pools. They had fascinated me since I was a child; they were one of the only things I ever looked forward to when I had to come to Forks."

Having lived within sight of the ocean in Los Angeles from the age of 4 to 14, tide pooling was a common childhood endeavor, and I still love to do it. What I called "jellyfish" as a child and when I wrote this column are now properly referred to as "sea jellies." Since they are not actually fish, the old term is problematic, which is the same reason that the newer term "sea stars" has replaced "starfish."

Animal Movie Mistakes
September 6, 2010

I'm not looking for realism at the movies. I understand that bullets only bounce off Superman's chest because he is imaginary, that having someone at "hello" only happens in the fictional *Jerry Maguire*, and that it takes the world's greatest minds and most advanced technology to create the worlds of *Avatar, Star Wars, Harry Potter,* and *Shrek*. What drives me nails-on-the-chalkboard, going-30-miles-an-hour-in-the-passing-lane nuts are errors about animals that are due to ignorance. Some inaccuracies about animals serve a purpose, and are therefore forgivable. Others are not so easy to forgive.

Forgivable Errors

In the movie *Finding Nemo,* the eyes of the fish are on the front of the faces. I found this distracting at first, since most fish eyes are positioned laterally, but then accepted it as a way for the artists to make the characters more human-like in their expressions and thus more accessible.

Similarly, the title character in *Bolt* has the whites of his eyes showing throughout the movie, which is not typical of real dogs. The purpose of this depiction, in addition to the fact that he erroneously lacked eyelashes and whiskers, was presumably to make him look more childlike and endearing.

When Mad-Eye Moody demonstrates the unforgiveable curses in the book *Harry Potter and the Goblet of*

Fire, it's a spider that is the victim of the Imperius, Cruciatus, and Killing Curses. In the movie, it's a tailless whip scorpion, which I can forgive because they are so extraordinarily cool looking that they fit the wizarding world.

In countless movies, an image of a Bald Eagle is accompanied by a Red-tailed Hawk call, which always makes me cringe. And yet, my logical self accepts this mistake because while the hawk call is a husky descending scream that sounds powerful, the high-pitched eagle call, though quite beautiful to many, sounds a bit like a squeaky toy to others.

Unforgivable Errors
The use of a Laughing Kookaburra call in the background of Tarzan movies to illustrate the wildness of the African jungle seems just plain wrong because this bird lives in Australia, not Africa.

Leafcutter ants only occur in the western hemisphere, which is why the opening scene of *The Lion King* showing leafcutter ants walking across the African Plains bothers me. I have no problem with meerkats and warthogs that sing, talking lions, or hyenas in a conspiracy with a scheming lion, but leafcutters in Africa upset my sense of reality.

The Antz is full of errors. Top offenders are that the legs are shown connected to the wrong part of the body and the film depicts males as soldiers when in fact only female ants function as soldiers in nature.

Bee Movie is similarly devoid of concern for scientific accuracy. The movie is about a young male bee's quest to choose his lifetime job within the colony. Actually, work in bee colonies is done almost exclusively by females, and they change jobs repeatedly as they age—a system known as "age polyethism."

I like being transported to imaginary worlds by movies. It's just that when they are based on the real world, I think certain similarities should carry over.

As much as I object to many animal-related errors in movies, I try not to let it ruin the movies for me. Luckily, I can vent to my biologist husband, who is as likely as I am to be troubled by these inaccuracies.

Crows Are Smart and Social

September 20, 2010

When a group of animals is referred to as a "murder" rather than as a pack, herd, school, cluster, band, troop, tribe, colony, swarm, mob, pride, covey, or flock, you have to wonder just how accurate it is to describe those animals as "social."

And yet, despite a group of them being referred to as a "murder of crows," these birds have some of the most complex social behavior known to science.

Crows roost in groups that can number in the thousands. They forage in groups as well, although these groups are much smaller. When a breeding pair is raising a new clutch, they often have helpers, usually birds from previous broods. Crows pass knowledge to each other so that something one crow has learned can be shared with others. For example, chemicals in ants can help eliminate crow parasites, so crows can get rid of parasites by lying on an anthill or by crushing ants into their feathers. This information is passed down from one generation of crows to the next.

It typically takes a certain mental capacity to handle the complexities of social life, so social animals tend to be smart. Crows are no exception. In fact, scientists consider them among the world's best problem solvers. For example, some crows drop hard-to-eat food such as clams and nuts with thick shells onto roadways so that cars will run over them and break open their otherwise impenetrable coverings. Then, the crows can feast on the food that otherwise would have been unavailable to them. In some cases, crows have even been reported to drop the food into crosswalks, and then retrieve it when the lights change to let pedestrians cross.

Crows use tools, which is considered by many to be the pinnacle of intelligence. Tool use was once considered a trait that distinguished people from animals, which is why Jane Goodall's observations of chimpanzees using tools caused such a brouhaha back in the 1960s. Along with a variety of other birds, primates, otters, elephants, and octopuses, crows are now known to use tools. Crows make at least two distinct types: they make hooks out of sticks and use them to remove grubs from holes in trees, and they turn stiff leaves into sharp implements that they use to sort through leaf litter looking for invertebrates to eat.

Aesop's Fable "The Crow and the Pitcher" shows that the perception of crows as highly intelligent is not new. In this story, a crow that was nearly dead from thirst finds a pitcher with very little water in it. His beak was too short to reach the water and after many attempts to get the water, he almost gives up. Then an idea occurs to him. One by one, he drops pebbles into the pitcher until the water level rises enough that he can reach it and quench his thirst. While scholars argue that the moral of the story may be that necessity is the mother of invention, that persistence is a virtue or that cleverness is more powerful than force, there is general agreement that crows' intelligence is well represented by the tale.

Watching crows is always worthwhile because they do so many interesting things! Their intelligence and problem-solving abilities continue to be explored by scientists.

The Display of the Peacock
October 18, 2010

Peacocks are famous for the grandeur of their courtship display, which involves roughly 200 feathers that grow up to 5 feet in length and are brilliantly colored with iridescent blue and green eyespots.

In fact, their display is so dramatic, and involves such enormous exaggeration of body parts, that evolutionary biologists from Darwin to those of the present day have constantly found themselves called upon to explain how such a feature could have evolved. The issue is that the excessive length of feathers and coloration are clearly disadvantageous for survival. They make the birds easier for predators to spot and slower in their attempts to escape—a bad combination.

It was Darwin who developed the explanatory theories of sexual selection and female mate choice. Evolution occurs because of the competition to reproduce and leave behind offspring. Individuals with random variations in their genes that make them more successful at producing offspring pass on those genes. One form of competition to reproduce involves competition for mates. It can take the form of combat where the winner acquires mates, or involve mate choice, which is most typically performed by the females.

Females of many species are choosy about which males they mate with. One theory suggests that females who choose showy males are choosing males who are healthy enough to produce such elaborate ornamentation and coloration, and that choosing healthy males is a good way to increase their chances of producing offspring with good genes.

During mating season, the males, who are called peacocks, congregate in groups called "leks." Females, or peahens, come to visit the leks. While the females move through the leks to assess their choices, the males display their feathers while hopping about and shaking. This mating dance is called "shivering."

Why the peacock's display evolved is a separate issue from how it evolved. The steps that led to the peacock's exaggerated features are well understood from comparisons among related species. The peacock's display is the most elaborate one out of a group of displays in the galliform, or poultry-like birds.

Male jungle fowl, and the domestic chickens that are descended from the jungle fowl, have a calling to food behavior in which the male attracts the female by bending forward and pointing to a food item on the ground with his beak. As he leans his body forward, his tail is raised naturally, but the tail position relative to his body is the same as in any other posture. In grouse, turkeys, and pheasants, the tails of the males have various degrees of coloration, patterning, and exaggerated development in size that are visible during the calling to food behavior and some species also exhibit spreading and fanning movements.

In the peacock, the courtship display has the highest degree of exaggeration in both the behavior and the feathers. This bird has been admired through the ages for the splendor of its display. The term "peacocking" refers to the practice of men dressing and behaving in an over-the-top way to attract women—a practice made possible by the fact that female mate choice occurs.

The elaborate tail of the peacock is often described in terms of the males' displays, but the whole system can only make sense when combined with a discussion of female mate choice.

Birds are Dinosaurs

November 1, 2010

In our house, it's pretty typical to see a child come around a corner impersonating a *Tyrannosaurus rex* while another child races by yelling, "You can't catch me! I'm a *Gallimimus*, so I'm faster!" At another point during the day, you might see these same kids sitting quietly together watching the birds at our feeders.

Since scientists have long believed that birds are dinosaurs, their dinosaur play and bird watching are related pursuits. The idea that birds are dinosaurs, though commonly known and accepted in the scientific world, is novel to many people. Perhaps most startling, this means that dinosaurs are not extinct, which is counter to what most of us have been taught.

Yet, birds being an extant (existing) lineage of the dinosaurs doesn't change the fact that the vast majority of dinosaur species are extinct, including well-known ones such as *Triceratops, Allosaurus, Apatosaurus, Hadrosaurus, Iguanodon, Velociraptor, Parasaurolophus, Stegosaurus,* and *Ankylosaurus*. It simply means that one branch of the dinosaur family tree survived to the present day, and the living members of that branch are our feathered friends. Specifically, birds evolved from theropod dinosaurs. Theropods include *Dilophosaurus, Coelophysis, Tyrannosaurus, Deinonychus, Ceratosaurus, Oviraptor,* and *Ornithomimus.*

The idea that birds descended from dinosaurs originated in the 1800s, when scientists noticed the many similarities in the skeletons of birds and dinosaurs. In the 1960s, scientist John Ostrom observed 22 features of the skeleton that occurred only in meat-eating dinosaurs and birds, and in no other animals. After additional studies by many researchers, the number of skeletal characteristics that occur only in these two groups is 85.

Further evidence of the evolutionary link between birds and extinct dinosaurs comes from comparing fossils along the lineage that includes *Archaeopteryx. Archaeopteryx* lived during the Jurassic period about 150 million years ago and is generally considered to be the first bird. It had many features in common with modern birds such as feathers, wings, a wishbone, and a partial reversal in the position of one toe. It also shared many traits with theropod dinosaurs such as teeth, clawed fingers, and a long bony tail.

Additional fossils that appear to be intermediate between earlier dinosaurs and modern birds have also been discovered. Important finds in many areas, particularly China and Spain, include fossils that are 30 or 40 million years more recent than *Archaeopteryx,* and are much more bird-like. For example, their bony tails are shorter and their claws are much reduced. Fossils found in China also include species that were clearly dinosaurs, but had feathers, which indicates an evolutionary link between these extinct dinosaurs and modern birds.

Another discovery that links birds and extinct dinosaurs relates to the evolution of the folding wrist bone. The structure of this joint allows birds to fold their wings fully while at rest and partially during the upstroke when flying, which greatly increases the efficiency of flight. Some dinosaur fossils from species that lived before the evolution of flight have a wrist bone with the same sort of flexible wrist, indicating the ability to fold their feathered arms.

Though most dinosaurs went extinct millions of years ago, it's exciting to think that a few still live among us.

The way that many people object to the idea that birds are dinosaurs reminds me of the way many people reject the statement that Pluto is not a planet. For many, it is hard to adjust to information that differs from childhood lessons, yet science only moves forward as we accept new information and reject old ideas. (And I say that as a person who lives in Flagstaff, Arizona, where Pluto was discovered, and where its demotion to dwarf planet was taken quite personally.)

Turkey Time

November 15, 2010

"How do you keep a turkey in suspense?" "I'll tell you later!" That was my official welcome to the Thanksgiving season. (My children love to tell jokes.) Around this time each year, most of us spend a lot of time considering how to prepare our turkeys for the Thanksgiving feast. A little less time is generally devoted to what turkeys are like while alive, which is too bad because there a lot of interesting things to know about them.

Wild and domesticated turkeys are very different. One of the biggest differences is that domesticated turkeys cannot fly, but wild ones can. Flight is not possible in farmed turkeys because they have been bred for greater weight and larger breast tissue than wild turkeys. The heaviest turkey ever raised weighed 86 pounds.

Wild turkeys can fly at speeds over 50 miles per hour, and run as fast as 25 miles per hour. Their field of vision is about 270 degrees and they can see movement 100 yards away.

A group of turkeys is called a flock. Males are toms, females are hens, and the babies are called poults. Another term in the turkey lexicon is "snood." The snood is the fleshy growth that starts at the base of the beak, and is long enough to hang over the beak in males.

Only tom turkeys gobble. They do so seasonally in spring and fall to attract mates, and typically do so when they hear loud noises or when they are preparing to roost at night. Toms have about 3,500 feathers at maturity.

Benjamin Franklin liked the idea of the turkey being our national bird rather than the eagle. This is known from a letter Franklin wrote to his daughter objecting to the eagle, but there is no evidence that Franklin ever actually proposed that the turkey be our national bird. Franklin considered the turkey to be typically American and he thought the eagle was of questionable moral character. Specifically, he objected to the eagle's habit of stealing food that other birds have caught.

Turkeys originated in Central and North America and have been around for about 10 million years. They came close to disappearing in the early 1900s, but now every state in the continental United States has them.

Turkeys have a few quirky attributes. For example, they have heart attacks. When the Air Force was breaking the sound barrier in test runs, turkeys nearby would drop dead. Their heads change from gray to brighter colors including blue, white, and red when they become excited or distressed. The idea that they can drown if they look up in the rain is simply a myth.

Though turkeys are amazing animals, most of us just enjoy them on the table, or through jokes. Here are a few I've heard:

What key has legs and can't open doors? Turkey.

Why shouldn't you let turkeys talk to your children? They have such fowl language.

Can a turkey jump higher than the Empire State Building? Yes, buildings can't jump at all.

Which side of the turkey has the most feathers? The outside.

What do you get when you cross a turkey with an octopus? Enough drumsticks for Thanksgiving dinner.

There are wild turkeys in Flagstaff, and we are usually able to get a few sightings each year. It's astounding how a group of dozens of these enormous birds can be almost invisible to us in the woods.

The Call of the Coqui
December 6, 2010

The sound of the calling frogs in the rain forest at night is magnificent. In the early hours of our last morning on a recent vacation in Puerto Rico, I listened with great attention to the calling of the coqui frogs, knowing that when the sun came up, I would have to say good-bye to this glorious chatter. The calling commences when the sun sets, and it continues throughout the night until first light.

The coqui is such a powerful symbol of identity in this beautiful commonwealth of the United States that it has given rise to the expression "más puertorriqueño que el coquí," which means "more Puerto Rican than the coqui."

Sixteen different coqui species live on the island, with 13 occurring in the Caribbean National Forest, which is a treasured 28,000-acre area of rain forest. The best known and most abundant species is the common coqui, *Eleutherodactylus coqui*. Since these frogs are small, anywhere from 1.5 to 8 cm long, it has been said that the frog's name is longer than its body.

The genus contains over 200 species, which some claim make it the vertebrate genus with the most species. *Eleutherodactylus* means "free toes" and refers to the fact that these frogs, unlike most frogs, do not have a membrane between their toes. Their toes have little pads on the tips, which allow them to hold onto leaves, walls, and other surfaces, but the lack of an interdigital membrane suggests that they are not adapted for swimming.

Coqui frogs are unusual in the world of amphibians in that they lack a free-living larval stage. That is to say, coqui frogs pass through their tadpole stage while still in the egg and when they emerge from the egg, they are fully formed and functional froglets, though very small. The males guard the eggs to prevent them from drying out and remain in the nest for several days after the new froglets appear. Females lay clutches of about 25 to 30 eggs and they do so roughly five times per year. Breeding occurs throughout the year in Puerto Rico, though the wet season is a particularly active time for coqui reproduction.

The frog's name is onomatopoeic for its "Ko-kee!" call, which is an advertisement call used by males during the mating season to attract females with which to mate. Only male coqui frogs call. The call is actually made up of two distinct parts—the "Ko" and the "kee" components of the call have very different purposes and are intended for different listeners. The "Ko" is a call to other males to tell them to stay away—sort of a "Back off mister, this spot is taken" kind of a call. The "kee" attracts females, so I would translate it loosely to mean, "Hey baby baby."

I like the call of the coqui enough that I was not surprised to learn that it has been made into a ringtone, though I doubt I find the call as pleasing as the average female of the species finds the real thing.

A professor of mine at UCLA and his graduate students studied the calls of *Eleutherodactylus coqui* frogs, so I knew a bit about them before our 2010 trip to Puerto Rico, and was so thrilled to hear the familiar call in the wild for the first time.

Where the Buffalo Roam

December 20, 2010

The animals on islands are typically those whose ancestors crossed the water barrier, so animals capable of swimming, drifting, or flying are well represented on islands worldwide.

Birds are often easily able to cross water barriers, as are insects. As the only mammals capable of powered flight, bats are among the most numerous mammals on islands. Whenever you are on an island and see an animal, it is interesting to consider how its ancestors arrived there from the mainland (or from another island). Sometimes, it's not obvious how that could have happened, as in the example of the buffalo that roam the hillsides of Catalina Island, which is a little more than 20 miles away from the Southern California mainland.

They didn't fly or swim there, nor did they drift. In actual fact, they were brought there for the sake of a movie. In 1924, 14 buffalo were brought to Catalina Island for the movie *The Vanishing American,* based on the novel by Zane Grey. Interestingly, Grey had a vacation home on Catalina Island. After the movie had been filmed, the buffalo were left loose on the island. Despite a few of them being shot and killed, enough were born to bring the population up to about 28 by 1934.

By 1969, the healthy, thriving population on the island numbered almost 400. Because extensive inbreeding was a major concern, 15 more buffalo were brought to the island from Wyoming. At that time, a program of annually culling the herd and periodically introducing new individuals to the population was begun. The herd has been as large as 600 buffalo, though they are currently being managed to maintain a population of about 150–200 individuals. Strategies for managing the herd to maintain its proper size involve a combination of birth control injections to female buffalo and the periodic relocation of some animals to other areas of the country, such as to the plains of South Dakota.

From an ecological perspective, having a population of animals that don't really belong on Catalina is a problem. They trample or eat many of the delicate plants that are so crucial for preventing erosion on this rocky island. They also knock over barriers that are intended to protect the threatened ironwood trees. However, they are considered part of the island's cultural fabric, which is perhaps why they have remained there rather than all being relocated to areas where buffalo naturally occur.

The Catalina Island Conservancy manages the buffalo herd and is eager to do so in a way that avoids a public relations nightmare. Starting in 1989, they hired sharpshooters to kill pigs and goats from helicopters in order to decrease the populations of these animals for ecological reasons similar to those that favor a reduction in the buffalo herd. No matter how good their intentions, the shooting of goats and pigs did not go over well, while the relocation and birth control measures for buffalo have been more favorably received. In fact, seldom is heard a discouraging word.

These large mammals are actually bison, not buffalo, but in the song, *Home on the Range*, they are called buffalo even though there are no native buffalo in North America, so I took artistic license when writing this column. (Pronghorn are referred to in the song as antelope—it's a nomenclature nightmare.) Also, since writing this column, it has been noted that the film *The Vanishing American* has no bison in it nor any landscapes that resemble Catalina Island. It is now believed that the bison were brought to the island for the filming of a 1925 film called *The Thundering Herd*.

Wing Shape Affects Flight
January 3, 2011

Birds fly using the same principles as airplanes. Although, since birds were flying millions of years before humans, it's more accurate to say that airplanes use the same principles as birds. Both birds and planes fly differently depending on the type of wings they have. Different wing shapes represent trade-offs between various desirable features of flight such as speed, energy efficiency, and maneuverability.

The variation in wing size and wing shape can be described by two main parameters: aspect ratio and wing loading. Aspect ratio is the square of the wingspan divided by the area of the wing. Wings that are long and narrow have a high aspect ratio while short, stout wings have a low aspect ratio. Birds whose wings have a low aspect ratio are better able to make sharp turns when flying than birds with high aspect ratio wings.

Wing loading is the weight of the bird divided by the area of the wing. Birds with low wing loading will achieve greater lift at a given speed than birds with high wing loading and need less power to sustain flight. Low wing loading is also associated with greater maneuverability. There are four basic types of wings and each one is best adapted for a different kind of flight.

Short, pointed wings. This sort of wing allows birds to fly at the highest speeds. The fastest bird is the Peregrine Falcon, which has this sort of high-speed wing. They have been recorded at speeds around 175 miles per hour while diving. Many other falcons have wings that are short and pointed, as do ducks and swifts. In fact, the Spine-tailed Swift holds the record for the fastest straight powered flight, having achieved speeds of 105 miles per hour.

Elliptical wings. These wings are short and round, which maximizes a bird's maneuverability. Raptors who live in forests and other places with relatively dense vegetation have elliptical wings, as do many non-migratory perching birds. Birds who often take to the air quickly to avoid predators, such as pheasants and partridges, also typically have elliptical wings.

High aspect ratio wings. Wings of this type are longer than they are wide and usually have low wing loading. Birds with such wings either fly slowly, or their flight involves gliding and soaring. Many seabirds have high aspect ratio wings. These birds require a long taxi period to become airborne, but are efficient once in flight. Most long-distance fliers have wings with a high aspect ratio, including the albatross, which is well known for its long-range flights.

High lift wings. These are also called soaring wings with deep slots. The gaps at the tips between the primary wing feathers lower turbulence at the wing tips and provide additional lift. Birds with these wings have enough lift to keep their large bodies in the air even when burdened by the extra weight of prey. Wings of this shape are seen in large birds such as storks, pelicans, eagles, and vultures.

Though usually interpreted as inspirational, William Henry Hudson expressed a simple truth when he wrote, "You cannot fly like an eagle with the wings of a wren."

Once you start to pay attention to the different shapes of bird wings, it's easy to see the variation and relate it the lifestyles and demands of flight in different species.

Courtship Gifts
February 7, 2011

It's tricky to distinguish an "Only Red Roses Will Do For Valentine's Day" type of person from an "Anything But Red Roses—They're So Cliché" type of person. Perhaps this explains why many men experience more fear when contemplating the choice of a Valentine's Day present than when thinking about public speaking or violent crime.

Interestingly, we are not the only species in which males give courtship presents, nor the only species in which females critique those gifts.

Male balloon flies give females a large white "balloon" made of silken thread. The females judge whether a male is worthy of her attentions based largely on the quality of this gift. Males who offer her nothing are rejected outright. Males who produce a small or otherwise inadequate balloon will only be given a very short time to mate with her, thereby fertilizing fewer eggs.

This silk balloon has no intrinsic value to the female. It's not food since she eats only nectar. She doesn't lay her eggs in it or use it for protection or for any other practical purpose. However, this balloon is very expensive for the male to make because it takes a lot of energy and time to produce.

The evolution of this particular gift proceeded in a step-by-step fashion. Presenting prey to females as a courtship gift is common among insects and very practical. If a male is a good hunter and can present her with a high-quality prey item, it is likely that he has good genes and is a worthy mate. Furthermore, the prey is useful because the amount of food a female eats is directly related to her ability to produce a lot of eggs and eggs of high quality.

Some relatives of the balloon flies, representing the next evolutionary step in the process, wrap the prey in a few threads of silk to immobilize it and make it easier to deliver and consume. There's an intermediate step between wrapping the prey in silk and making an empty balloon: In some species, a male wraps a piece of vegetation in the balloon. The vegetation is of no value to the female.

Finally, the most advanced stage in this evolutionary sequence is the production of a highly costly (in terms of time and energy) but completely useless gift, the quality of which is a major factor in a female balloon fly's decision about whether or not to mate with a male.

Expensive gifts that matter to females and play a role in which males they choose as mates but have no practical value are not unique to balloon flies. Of course, part of me thinks that spinning a hollow balloon of silk is no less crazy than trying to get millions of roses to bloom in North America in the middle of February or thinking that the most prized gift of all is a colorless rock that forms when carbon is compressed by tons of force at high temperatures, but that's just me.

Here's hoping that everyone's gifts are thoughtfully chosen and deeply appreciated. Happy Valentine's Day to flies and people alike.

Useless gifts that are nevertheless highly valued and used to assess a potential mate's value in our species (I'm looking at you, diamond rings!) make more sense when we see the evolutionary parallels in other species.

GloFish® a Trendy Pet
February 21, 2011

I overheard a pet store employee describing GloFish® as zebra danios that have been crossed with jellyfish, which sounded so unbelievably cool to this science geek that I didn't believe it. And in fact, it's not technically true. While GloFish® are zebra danios and some of them have genes from a jellyfish, they are not a cross of the two species.

Rather, GloFish® are genetically modified organisms, meaning that they are organisms whose genetic material has been altered using recombinant DNA technology. These techniques consist of combining the DNA from more than one organism into a single packet and then transferring it into an organism. The color comes from either a jellyfish gene or from a gene in sea coral. Because GloFish® have DNA from more than one species, they are actually a type of genetically modified organism called a transgenic organism.

Genetically modified foods have been on the market since the early 1990s, but it's only been since 2004 that you could purchase a genetically modified pet. That's when GloFish® first became available.

The GloFish® is a trademarked fish, which was a wise move from a marketing standpoint since these fish are selling like Harry Potter books. The reason they are so popular is not because of the amazing science—specifically the use of genetic engineering—used to produce them. Rather, what's so desirable about them in most people's minds is that they are bright fluorescent colors, making the average aquarium look like a pack of highlighters has just exploded or started to breed within. There are a lot of different colors—Yellow®, Starfire Red®, Electric Green®, and SunBurst Orange®.

Creating trendy pets was not the goal of the scientists who developed them. Rather, they were attempting to create a fish that only fluoresced in the presence of certain toxins, so that the fish could be used as an indicator of environmental pollution. Developing a fish that was constantly bioluminescent in both natural and UV light was just the first step toward this goal.

Scientists developed these fish in 1999 from zebra danios, which are common laboratory research organisms. Zebra danios are tropical fish whose native environment is rivers in India and Bangladesh. They are 3 centimeters long and typically have blue and gold stripes. Though hundreds of millions of them have been sold as pets in the last half century in the United States, no populations of them have ever become established in the wild here, most likely because tropical fish are unable to survive our temperate climate.

Now GloFish® are big in the pet trade, and one concern is the risk of them establishing wild populations. This concern is perhaps the reason that GloFish® are not allowed in Canada, in the European Union, or in California. They are not allowed in California because of a regulation restricting all genetically modified fish. This statute pre-dates GloFish® and reflects concerns about a fast-growing genetically modified salmon.

GloFish® have sparked a worldwide controversy. There are those who call them "FrankenFish" and express concern about them because they are genetically modified, but there are others who proclaim, "Let there be GloFish®."

More than a decade after I wrote this column, GloFish® remain popular and controversial.

Why Do Cats Play with Their Food?

March 14, 2011

Most parents repeatedly tell their kids not to play with their food, but they do it anyway. Similarly, a large number of cats play with their food though it appalls most owners to see their cat apparently torturing a bird or small rodent in what seems like a cruel game.

Actually, the way that cats let go of and then recapture their prey is not a way for them to have fun, but rather a way for the cats to protect themselves from serious injury. Cats kill their prey by breaking the spinal cord with a strong bite to the neck. If a cat must let go of the animal in order to grab it on the neck, that cat is risking escape or retaliation by their prey.

The prey that cats hunt have weapons of their own and a cat can easily be injured by them. For an animal such as a cat with a short muzzle to reach the neck of an animal, even one that is already caught, is to risk injury to the eyes or face. Rodents such as mice and rats can bite and scratch, and a bird can inflict a lot of damage by either biting or pecking with a sharp beak. So cats tire out their prey before making a killing bite in order to minimize their own risk of injury.

A study done decades ago investigated the playful behavior that cats often exhibit with their prey. The scientist Maxeen Biben presented cats with different sizes of prey. The small prey were mice and the large prey were rats. Some cats were hungry, others were very hungry, and some cats had recently eaten. Biben found that when cats were given a rat, they were more likely to play with it if they were very hungry than if they had just eaten or were only a little hungry. She suggested that cats have to be very hungry to attempt to kill large prey such as a rat, and that they must perform these playful behaviors in order to be able to make the kill safely. A large prey animal such as a rat is even more likely than a small animal like a mouse to seriously injure the cat with scratches or bites, either of which can become infected and result in the cat's death.

Another finding in Biben's study is that cats played with their small prey less when they were hungrier. Cats who had recently eaten tended to play with their mice prey longer, only killing it once the prey had become so tired that the cat could easily kill it safely. Hungrier cats played for shorter periods of time before killing mice. Perhaps when they were hungrier, they were more willing to risk injury, since the benefit of killing more quickly was being able to eat sooner.

The next time you notice your children playing with their food, recast their behavior as "catlike" and applaud their excellent self-preservation skills.

The way that cats play with toys often mimics their hunting behavior, but it looks less playful when put in the context of predatory behavior.

Scared of Spiders?

April 4, 2011

Ron Weasley has made being frightened of spiders respectable because this Harry Potter character is plenty cool despite his phobia of arachnids. Being afraid of spiders is common, but I only know one person who is truly phobic of them.

I met Sarah in Venezuela when we were both doing field research on animal behavior for our graduate degrees. The way I found out about her extreme fear of spiders was dangerous to both of us.

We were walking through a dense, lush area of the tropical forest searching for a troop of wedge-capped capuchin monkeys for her to observe for the day. The path was narrow, overgrown and a challenge to follow. Sarah mentioned to me ahead of time that she did not care for spiders, which is why she likes to walk slowly and attempt to break the spider webs across the path before running into them. Once we began our walk, I learned that the way she did this was to carry her machete in front of her face as we walked. I thought it seemed like a good system. It was not.

A few minutes after we had turned onto a trail even more overgrown than the first one, and far out of site of the main path, Sarah's machete and her hand hit a spider web. The spider swung by her ear on a fragment of its web. The next thing to fly by me was Sarah's machete, which she had tossed high into the air with a scream that robbed me of the upper register of my hearing for several hours. The machete sailing high into the air was not technically a problem, but the return flight to earth involved a near miss of my shoulder and then it landed within inches of my foot.

I told her that henceforth, I would be happy to lead. True, I dislike the feeling of a spider web clinging to my face, and I wasn't overjoyed about the prospect of having an unidentified tropical spider on me either, but I considered both situations highly preferable to dealing with machetes in full attack mode at close range.

Though Sarah is clearly arachnophobic, there is another group of people whose feelings toward spiders are currently at least as negative as hers, and perhaps even more so: executives of the Mazda Car Company, who recently had to recall over 65,000 cars because they were being invaded by spiders.

Specifically, yellow sac spiders were building webs in the vent lines of Mazda6 cars. These spiders like the smell of gasoline, and the Mazda6 has a ventilation system that lets out just enough gasoline to attract the spiders without causing them to die. If the spiders' webs completely block these crucial venting paths, there is a risk that pressure could build up enough to cause cracking in the fuel tank, resulting in leakage and the possibility of a fire.

It's a good thing Sarah doesn't drive a Mazda. Throwing machetes while driving would constitute a safety hazard.

Many people are afraid of spiders, but that has not kept them from becoming the new "it" animal for the scientific study of animal behavior. Research on these animals has become increasingly popular and anyone attending a scientific conference about animal behavior will be treated to many great talks about them.

The Dance of the Honey Bees
April 18, 2011

It's in Chaucer's *Canterbury Tales,* written in the late 1300s, that we first find the expression "busy as bees," though now we usually use the singular, "busy as a bee." Bees have always been considered industrious workers, and rigorous study has confirmed this long-standing belief.

According to Tom Seeley, author of *Honeybee Ecology: A Study of Adaptation in Social Life,* each year an average hive raises 150,000 workers who collectively go on several million foraging trips, some of which are as far as 10 kilometers from the hive, and travel 20 million kilometers. They eat 20 kilograms of pollen and 60 kilograms of honey.

Honey bees are efficient and effective foragers. The nectar and pollen that fuel a beehive are scattered widely across the landscape in flowers whose offerings do not last long. Finding these resources and collecting the pollen and nectar quickly, before some other animal does, is the competitive game bees regularly win.

A bee who has discovered a good patch of flowers moves around in the hive in a characteristic way that scientists have called the "dance of the honey bees." Bees use this dance to communicate the location of a good food source to other members of the hive, allowing large numbers of bees to find it more quickly than if they each had to search independently.

A dancing bee walks in a straight line while wiggling her abdomen back and forth quickly. This straight walk while wiggling the abdomen is called the "waggle run." The bee then stops wiggling her abdomen and loops around in a circular path to the starting point of her waggle run. The bee repeatedly does a waggle run along the same straight path, then walks in a circular path back to her starting point.

Nobel Laureate Karl von Frisch figured out the meaning of the dance, in which the information about the direction and distance of the food source is symbolically encoded in the waggle run. The duration of the waggle run indicates the distance from the hive to the food. The longer the bee waggles, the farther the food source is away from the hive.

The direction to the food source is represented by the angle of the straight waggle run relative to the top of the hive. (Honey bee combs are vertical so that bees are clinging to the sides of them as they do all the work in the hive.) This angle corresponds to the angle of the food source from the azimuth of the sun, which is a line drawn from the sun to the horizon.

Astoundingly, bees adjust their dance as the reference point (the sun's azimuth) changes throughout the day. Bees alter the angle of their waggle runs (even if they have remained in the hive and therefore been unable to see the sun) to account for the change in this reference point.

If Chaucer had known more about the lives of bees, he might have used phrases like "as time-sensitive as bees" or even "as symbolic as bees" since they communicate with symbolic language, which was once thought to be done only by humans. Aah, the poetry we've missed because of a lack of knowledge.

The amount of information coded in the honey bee waggle dance is so astounding that many people refuse to believe it is possible for insects to communicate in such a complex way.

Mimicry in the Animal Kingdom
May 2, 2011

As soon as I noticed the tiny snake crawling over the toe of my tennis shoe, I grabbed my field notebook and recorded the pattern of yellow, red, and black stripes on its body. I wanted to know whether this snake was a coral snake or just a snake that mimicked one.

If it was a real coral snake and it bit me, I would be injected with a neurotoxin that can result in respiratory paralysis and is sometimes fatal. If it was a snake that looked like a coral snake but was in fact harmless, the worst I would face in the unlikely event that it bit me was some irritation.

Though I love snakes and enjoy holding them, I choose never to handle venomous ones. This snake avoided being picked up and otherwise harassed despite my curiosity about it, because its color pattern left me uncertain about whether it was a harmless snake or a venomous one.

Harmless animals that closely resemble poisonous or venomous ones are common in the animal world. There are flies that look like bees, non-toxic butterflies that are hard to distinguish from toxic ones, and non-venomous milk and king snakes that share the striking color patterns of deadly coral snakes. The harmless animals benefit by mimicking the dangerous ones because predators tend to avoid species that are poisonous or venomous and avoid the mimics because they are hard to distinguish from those that are hazardous.

This sort of mimicry is called "Batesian mimicry" in honor of the English naturalist and explorer Henry Walter Bates who first described the phenomenon. He spent many years traveling in South America, observing and collecting thousands of species that were as yet unknown to science.

Bates first described the mimicry that now bears his name with the example of the monarch and viceroy butterflies. The monarch butterfly's body is full of nasty tasting chemicals that can cause vomiting and even death in animals that eat them. It was long thought that the similarly colored viceroy was a palatable mimic, but further study revealed that the viceroy is actually poisonous, too. So ironically, this system is no longer considered an example of Batesian mimicry at all.

Mimics are never identical to the poisonous animals they resemble, but they can be hard to distinguish even upon close examination. The children's rhyme, "Red touches yellow, dangerous fellow; red touches black, friend to Jack," would not work to tell me whether the snake that I saw was venomous or not because this rhyme only refers to snakes in North America, and I was in Costa Rica at the time. In Central America, there are coral snakes where the red stripe is next to the black one, so relying on this rhyme instead of a proper field guide has caused many travelers to be bitten by a dangerous snake that they incorrectly thought was incapable of hurting them.

As it happens, the slithering youngster that invaded my personal space was *not* a coral snake. This harmless creature was protected from me, despite its lack of venom, by the effective defense mechanism known by the name of Batesian mimicry.

The evolution of Batesian mimicry has occurred across many taxonomic groups, including butterflies, beetles, flies, moths, frogs, and snakes.

Charging Elephants
June 20, 2011

Every stage in life has its standard tales. College students have stories of traveling through Europe sleeping in train stations. New parents tell of sleepless nights with screaming babies. Grandparents give details about the world's smartest kid, who just happens to be their grandchild. Similarly, biology graduate students doing field work in Africa have stories about charging elephants.

Elephants are a force to be reckoned with. Large ones are 13 feet at the shoulder and 13,000 pounds. Their foot-long molars weigh roughly 11 pounds. Big tusks reach 8 feet in length and weigh nearly 100 pounds. These facts hardly make the idea of being charged by them appealing, nor does the fact that they can run 25 miles per hour.

A friend from my graduate program was on safari in South Africa. As the jeep he and other scientists were in rounded a corner, they were thrilled to find themselves within a herd of elephants. That's when they were first charged by a large bull elephant coming from the side and crashing through the brush. The driver alertly sped forward to get away from the danger it presented. From the back seat came some worrisome advice, "When the elephant knocks the vehicle on its side and starts stomping it flat, you'll want to scream, but don't, it will only aggravate him." Then, as they raced forward, they saw another elephant ahead of them and that one was charging, too. So they backed up, and as they did so, they saw one coming toward them from behind. My friend asked the backseat expert, "What do we do now?" and received the answer, "Now we pray."

Another friend of mine was in Kenya studying patas monkeys. Adult monkeys give alarm calls whenever they see big cats, snakes, jackals, or eagles, but know not to do so in response to elephants, which pose no danger to them. When the baby monkeys alarm call, they often make mistakes, erroneously reacting when they see zebras, ostriches, and buffalo, as well as elephants. On one occasion when my friend heard a baby patas monkey give an alarm call, the other monkeys looked around but then did nothing, so she knew it was an error and thought little of it.

Unfortunately for my friend, the error was actually in response to a huge elephant that was probably 60 years old. He was with a group of females who were as surprised as my friend was to find themselves in such close proximity, which makes them panicky, and therefore dangerous. Also, males are at their most unpredictable when with a female group and more likely to charge, injure, and kill people. She and her field assistants managed to race to their car in time to avoid the charging elephant, but it was a close call.

Just as near misses with elephants are common, so is the reaction by fellow biologists upon hearing the stories. The standard question is not "Oh my goodness, are you okay?" but rather, "Did you get pictures?" It's just part of the standard experience of being charged by elephants.

On a recent safari in Tanzania, our family noted that our guide was not distressed by giant snakes, a cheetah that jumped up on our jeep, or nearby lions but was very nervous when elephants got close to our vehicle.

Bald Eagles
July 4, 2011

When the economist E. F. Schumacher said, "Eagles come in all shapes and sizes, but you will recognize them chiefly by their attitudes," he summed up the general impression of eagles, which is overwhelmingly positive. In fact, these birds were so highly regarded by our country's founders that the Bald Eagle was chosen to be the national emblem of the new nation that celebrates its 235th birthday today. Their long life, great strength, and majestic looks are often cited as the specific reasons behind this decision. They were also chosen because they occur only in the Americas, not counting a couple of sightings of vagrants in Ireland.

The Latin name, *Haliaeetus leucocephalus*, means white-headed sea eagle, and though seven other species occur elsewhere in the world, the Bald Eagle is the only sea eagle that lives in the Americas. Their Latin name is accurately descriptive, though their common name misses the mark. The heads of Bald Eagles are covered with feathers, and therefore not actually bald. Their name is derived from the Middle English word "balled" which does not refer to hairless, featherless skin, but rather means "white." Previously, this bird was known as the American Eagle and the White-headed Eagle.

One of the most easily recognized birds, the Bald Eagle is "all field mark," to quote Roger Tory Peterson (the famous naturalist and writer of field guides). That is, they are easy to identify because the white head and tail contrasted with the dark body is a unique and easy-to-see combination, at least if the bird is old enough to have developed its adult plumage. The white head and tail are present in birds that have reached the age of 4–5 years.

Part of what makes spotting Bald Eagles easy is their size. The females, which are larger than males, can have wingspans of up to 8 feet, reach heights over 3 feet, and weigh as much as 14 pounds.

Bald Eagles mainly eat fish, but they also eat ducks, carrion, rodents, and snakes. In order to hunt, these birds swoop down over the water and grab fish with their talons. Their vision is four times as sharp as a person with "perfect" vision and this excellent eyesight partially accounts for their success at catching fish. Bald Eagles can see a fish under the water from hundreds of feet above the surface. Because fish are such a large part of their diet, Bald Eagles are frequently spotted near water.

In fact, an excellent place to see Bald Eagles in Flagstaff is around Lake Mary and the surrounding hills, where many wintering Bald Eagles spend their time. It makes sense that we have Bald Eagles in Arizona, because it is such a common destination for snowbirds. Like many of our northern neighbors, Bald Eagles like to head south for the winter, and they are welcome additions to our area.

Quotes about the grandeur and beauty of eagles abound. Still, for memorability, you can't beat actor John Benfield's remark: "Eagles fly but weasels don't get sucked into jet engines."

We like observing Bald Eagles around town, especially the juveniles. Though we enjoy watching these birds of prey for their own sake, that's not the only reason for our interest. Juvenile Bald Eagles have dark heads and can resemble Golden Eagles, a bird that's far less common here. Usually, our suspicion that it's a Golden Eagle is just wishful thinking, but occasionally, we're treated to a sighting of one.

Small but Mighty
August 1, 2011

Consider all the animals whose weight equals two paperclips, two Cheez-It® crackers, or half a teaspoon of sugar. Only one species of bird in the world is that tiny—the Bee Hummingbird of Cuba. It weighs less than 2 grams and is 5 centimeters in length. The Giant Hummingbird of South America is the largest hummingbird, but is still a small bird at roughly 20 grams and just over 20 centimeters.

Little but athletic, hummingbirds are the only birds capable of flying both forward and backward. They can hover, go sideways, and fly upside down. Their flying abilities are incredible, in large part because of how fast they beat their wings. The rate is about 70 times per second when flying at their average speed of 25–30 miles per hour, but can increase to 200 times per second in a dive, during which they can attain speeds of 60 miles per hour. The humming noise created by such fast wing flapping accounts for their name.

To fly as they do requires great strength, which explains why their flight muscles make up 30 percent of their weight. For comparison, the corresponding human pectoral muscles are about 5 percent of our body weight.

In addition to power, these feats of flight are possible because of a rapid metabolism. Their hearts beat about 250 times per minute at rest and up to 1,260 times per minute during exertion. They take about 250 breaths each minute while at rest and maintain a body temperature of about 107 degrees F. This level of energetic output means that hummingbirds must eat a lot. They can eat eight times their body weight in a day, which in human terms translates to a 175-pound man eating 1,400 pounds of food in a single day or a 130-pound woman eating 1,040 pounds. Hummingbirds visit an average of 1,000 flowers per day for nectar, lapping it up at about 13 licks per second. In addition to eating nectar, they consume insects, which is how they get their protein.

Perhaps it's because of their high energetic demands that hummingbirds sleep at night in a state of torpor (hibernation-like state), which conserves energy. Their heart rate drops to 50 beats per minute and they can lower their body temperature to below 40 degrees F. To the casual observer, they seem dead, sometimes even hanging upside down. It takes an hour for them to fully recover from torpor, which can be fatal to a weak hummer.

Their normal energy demands are high but increase prior to migration. It's common for them to double their body weight before migratory flights. Some species migrate thousands of miles twice each year. Many hummingbirds winter in the southern United States, Mexico, and Central America, but spend the breeding season in the northern continental United States, Canada, and Alaska. Some individuals fly the 500-mile stretch over the Gulf of Mexico without stopping, which takes almost 20 hours, while others stay along the Texas coast.

These amazing creatures support the motto that good things come in small packages.

Since hummingbirds occur only in the New World, many visitors from Africa, Asia, and Europe are thrilled when they travel here and can see them. In the eastern United States, there is only one species (two in some southern areas), so those of us who live in the west are very fortunate indeed, especially in Arizona, which has over a dozen species—six of which we saw just in our back yard last year!

Living on a Hamster Wheel
August 22, 2011

Tearing down the staircase to retrieve part of my dissertation from the printer in the main office, I flew by the chair of the department, who was racing up the stairs two at a time. A few minutes later, as I was running back up three flights of stairs to print the few pages that had been numbered incorrectly, this same professor was running down the stairs at a speed far surpassing basic safety precautions. Twice more we passed like this, until finally he said, "We seem to be stuck on the same hamster wheel."

Real hamsters often spend considerable time running on their wheels. In the wild, they run many miles each night searching for food. They need a lot of exercise when kept as pets, which is why hamster wheels are such an important piece of equipment. Most pet hamsters love to run on their wheels, and cover around 5 miles in a typical night. That is why a high-quality, quiet wheel is worth the extra money, especially if the hamster shares a room with any of the human members of the household. The best hamster wheels have a solid surface rather than rungs, which can cause injuries.

Runabout balls are an additional way to provide exercise, but you have to be careful lest your hamster overheat in these small plastic balls. Typically, 15–20 minutes is the maximum time an active hamster can be in one.

Although it's important, there is more to caring for hamsters than providing plenty of exercise. The first general rule is to leave them alone during the day, which is when these nocturnal animals sleep. Hamsters have a reputation for biting, but much of this undesirable use of their teeth is a result of bothering them while they are sleeping.

Speaking of their teeth, hamster's front teeth grow throughout their lives, so they require healthy options for items to chew on. Treat sticks and other chew toys should be available to them constantly.

Besides chewing, hamsters love burrowing into soft material. Many will shred paper and cotton to create an area for sleeping, but providing aspen or pine bedding is another option. Cedar shavings are not recommended because they may irritate a hamster's lungs and cause breathing problems. Weekly cleaning is advised as is completely changing the bedding monthly.

Hiding is common hamster behavior, so small cardboard boxes or even paper towel rolls make nice additions to your hamster's home. In addition to hiding in these structures, they sometimes play with them, along with other toys such as bird ladders and sticks.

Providing fresh food and water daily is important. Hamsters should be fed food that is meant specifically for them rather than for other types of pets. They should also be given small amounts of fresh fruits and vegetables such as bananas, apples, and green beans. Remove what the hamster doesn't eat so it won't rot.

Whether or not your own feelings of being on a hamster wheel allow you to empathize with your pet, it's kind to provide for them in a way that assures them a high quality of life—wheels and all.

I had a hamster as a teenager whose name was Mittens. He enjoyed his hamster wheel and his runabout ball.

Snakes Are Cool and Alarming

September 5, 2011

Traveling down a river in the Llanos region of South America, I felt apprehensive. I had accepted an invitation to join two Venezuelan scientists who were planning to capture and measure an anaconda that had been seen under the bridge.

I have strong, but conflicting views about snakes. To me, they are very cool, and also very alarming. Finding reasons to be alarmed is not hard. It is thought that the word "anaconda" comes from the word *anaikolra*, which in the Tamil language of India means "elephant killer." (Early Spanish settlers called this snake *matatoro* or "bull killer.") When I expressed my concern about trying to help catch an anaconda, my herpetologist friends just kept telling me that the anaconda is not venomous, which I already knew. I also knew that a snake capable of constricting a deer, tapir, or caiman would have little trouble dealing with me.

Of course, it's equally easy to be impressed by how cool snakes are and, after focusing on how lucky I was to be included, I was prepared for the adventure with the anaconda. I was not prepared, though, for the eyelash viper incident. We had stopped the boat and were enjoying the tropical forest by watching the capuchin monkeys in the canopy and listening to the macaws making a racket. Then, the man driving the boat said calmly in Spanish, "We need to move. There's an eyelash viper right above us." Thinking I had misunderstood my second language, I nonetheless looked up.

What I saw was a yellowish green snake that appeared to have eyelashes and was curled around a branch a foot over our heads. Eyelash vipers are beautiful snakes with the modified scales over the eyes giving rise to the common name, but it has the wickedly painful and dangerous bite typical of venomous pit vipers. Many people who are bitten by these snakes are doing exactly what we were doing—hanging out under a snake without being aware of its presence. Perhaps this is when I should have been alarmed, but all I could think when seeing this gorgeous creature at such close range was, "Cool!" Luckily, we got away safely, leaving us free to enjoy the memory unmarred by a medical emergency.

After the eyelash viper sighting, we proceeded to the spot where the anaconda had been spotted and went to work. It was exciting to see her head emerge from the water, pulled with a snake-catching pole, which is basically a long stick with a metal loop that tightens around the snake's neck enough to hold her but not enough to cause injury. As we pulled the snake out of the water, I could not help but be amazed at how her body seemed to go on and on. Helping to haul a nearly 20-foot snake weighing hundreds of pounds out of a murky river is not something everyone has done, and I consider myself lucky to be in the club.

The dangers faced that day were alarming, but these once-in-a-lifetime experiences with an anaconda and an eyelash viper were too cool for me to care.

The Venezuelan scientists were extremely excited about their experience and regularly returned to this spot of the river to see the snake they called Anna Anaconda.

What Is the Purpose of Play?

October 3, 2011

"The aging process has you firmly in its grasp if you never get the urge to throw a snowball," says Doug Larson.

Whether or not we remain playful as we grow older, having a pet cat, rat, bird, or dog makes it easy to observe play on a daily basis. Though play is a well-known phenomenon that occurs in many species, there's no consensus among experts on the purpose of this apparently pointless behavior that is costly in terms of energy, time, and the risk of injury.

There's no shortage of theories. Ideas about play's purpose include that play allows animals to use up surplus energy, or that it is strictly for fun or pleasure. It has been proposed that play is a way for animals to get exercise, or that it has a social function to help individuals work out their relationships in a non-serious context. Play might increase an animal's arousal level, which leads to more exploratory behavior and thus more exposure and learning opportunities in the animal's environment. A popular theory is that play provides animals an opportunity to practice behaviors that they must perfect later in life such as predatory behavior or courtship.

One specific idea among the "It's for practice" theories takes its cue from the old adage that war (or sailing, or flying) consists of "long periods of boredom punctuated by moments of sheer terror." It is during these moments of sheer terror—those rare, dangerous, and unexpected events—that the consequences of mistakes can be harmful or even fatal. This particular practice theory asserts that (1) Reacting appropriately to rare, dangerous and unexpected events is a matter of life and death, and (2) Opportunities to gain experience that increase the chances of survival in the face of such events are valuable enough to offset the costs of play.

The theory emphasizes that play provides this type of experience with a low-risk way to practice dealing with the unexpected. Play allows animals to develop physical and emotional responses to unexpected events that result in suddenly experiencing a loss of control. Proponents assert that the function of play is to provide animals with opportunities to increase the versatility of movements that are necessary for recovering from loss of balance, falling over, and other sudden shocks. They also propose that play functions to improve animals' emotional abilities to cope with unexpected stresses. They suggest that in order to train for the unexpected during play, animals actively create and seek out situations that mimic those that unexpected events can precipitate. They deliberately relax control over their own movements or put themselves into positions that are not advantageous.

This theory explains why play consists of rapid change between sequences of behavior with controlled movements much like those used in other situations and movements that cause a temporary loss of control. This type of switching between forms of behavior is cognitively demanding and results in a complex emotional state that is generally referred to as "having fun."

No matter what the greater purpose, having fun is reason enough to play. So go ahead . . . throw that snowball.

I'm pleased to report that I still enjoy snowball fights, so even though I am aging, not all hope is lost.

The Sting of Defense
October 17, 2011

"Cet animal est très méchant. Quand on l'attaque il se défend," is a quote from "La Ménagerie," a burlesque song from 1868. The translation is, "This animal is very bad. When attacked it defends itself," which pretty much sums up how most people feel about everything from snake bites to jellyfish stings.

People typically object to a bear charging and attacking when it has been startled or feel enraged when a scorpion stings the bare foot that stepped on it. Naturally, pain and fear have a tendency to make us angry and to prompt us to feel that what was done to us was not right or not fair. I dislike being hurt or scared as much as anyone, so I understand how the thought, "How dare they?" comes to mind.

I don't think of it that way, though. When I learn about the defensive behavior of animals, and even when I'm on the receiving end of it, I am thrilled and awed by what they can do (though sometimes pretty upset, too). Defense is the area of animal behavior that most fascinates me.

Consider a defense that most people despise—those hateful wasp stings. I know a thing or two about both the pain and wonder of their stings since the title of my doctoral dissertation was "Defensive Behavior in Tropical Social Wasps." Yes, I got stung a lot during my research, and yes, it hurt. No, I wasn't surprised that I was stung, and no, I didn't always take it gracefully. I yelled and said things that don't make me proud. I dropped my field notebook and any other supplies I was holding and generally made an idiot of myself.

But I was always intrigued by so many aspects of the defense of stinging wasps, even when I was still reeling from the pain. For example, only female wasps sting. Males do not have this defense, so people who say, "He stung me!" are talking nonsense, though to call them on this detail while they are suffering from the pain of a sting is tactless in the extreme. (Do mention it later though—education is never a waste!)

A honey bee can sting only one time, and then she dies. Wasps, though, can sting you over and over. Horrible, you say? I prefer the term "remarkable."

A true marvel of the sting as a defense mechanism is that the level of pain is completely out of proportion to the actual harm caused. The extreme pain of the sting suggests serious damage. However, except in the case of allergic reactions (which can cause anaphylactic shock, a true medical emergency that can even be fatal), the injury is actually minor. Sure, the pain is intense, but the damage is typically no more than a little redness and swelling.

So wasps are able to defend themselves because stings cause pain rather than because stings cause injuries which result in pain. It's an example of a deception that's incredible if you think about it, however painful it might be if you experience it.

The quote at the beginning of this column also served as the epigraph of my PhD dissertation.

Umwelt: The Perceptual World

November 7, 2011

My husband and I long ago learned to embrace the moments when our dog Bugsy showed his lack of Einsteinian brain powers.

On our farm in Wisconsin, he once suddenly began sniffing his own footsteps in the snow, backtracking for several hundred feet. I chuckled to myself, and even said aloud (to no effect), "Those are your own footsteps, you nut." Finally, he veered away from his own path and continued to track. Upon investigation, I realized that he was tracking a rabbit, and that he had first caught the trail from his own paw print. He had stepped onto a rabbit paw print hundreds of feet before and taken dozens of steps in the snow before catching the trail, and yet there was enough scent to get his attention and for him to follow. It made me wonder how many other times I erroneously identified his behavior as resulting from low IQ rather than existing in a different sensory world.

It's natural to assume that what we can sense is what's out there, but each species has a very different view of "what's out there." Every species has its own perceptual world, which is called the species' Umwelt. The most common translation for Umwelt is "subjective universe." Jakob von Uexküll came up with this term in

1907 to describe the phenomenon of organisms having different sensory experiences (even if they live in the same environment) because of varying capabilities of perception.

Humans are immersed in mainly visual and auditory stimuli, but our dogs, with their extraordinary olfaction abilities, are truly in a different world. Even within the same sensory modality, animals are capable of perceiving different things. For example, honey bees and humans both have excellent vision, but they see different parts of the color spectrum. Humans and bees both see the colors orange, yellow, green, blue, and purple. Honey bees, like most insects, are unable to perceive red, but they can see into the ultraviolet range, which humans can't see.

Additionally, bees are among the insects that can detect polarized light. That means that they can perceive not just the wavelength of light, which gives its specific color, but also the direction of the oscillation of the light wave, which can be in any direction along the 360 degrees perpendicular to its direction of travel. Polarized light is not part of humans' subjective universe, but bees see it and use the polarization pattern of the sky for navigation.

In order to develop a deep understanding of any animal, it is critical to understand the concept of Umwelt because animals' perceptual world determines what stimuli they tune into and which ones they do not attend to at all. From an overwhelming array of stimuli that reaches the body, only a very small percentage is picked up by the sense organs and transmitted to the brain. It is in the brain that this small fraction of stimuli is organized, analyzed, and interpreted to create for us the world in which we live. In a literal sense, the mind creates the world in which each of us lives.

I think it's exciting to realize that all species experience the world differently because it's a reminder that there's so much more out there in our world than we humans, being just one species of many, realize.

Honeyguides, Humans Work Together

November 21, 2011

The bird began to make a chattering call, a signal that is only used when communicating with humans. In addition to using auditory signals to get the person's attention, the bird flared its tail feathers, revealing noticeable white spots. Usually, animals signal to their own species, so this communication from bird to human is unusual. Perhaps even more unusual is the mutually beneficial relationship between these two species. The people in sub-Saharan Africa and this bird, the Greater Honeyguide, work together, with both species using their own strengths to help each other achieve what would be difficult or impossible alone.

The task they seek to accomplish is to feast on the rich resources of a beehive. The honey, honeycomb, and larvae inside a beehive are a bonanza for members of both species, and by working together they are able to exploit this resource.

The bird's expertise lies in finding the beehive in the first place. Though humans are capable of finding hives on their own, the bird's help decreases their search time by two-thirds. The Greater Honeyguide's skill at finding hives accounts for its scientific name, which is *Indicator indicator*.

A bird that knows the location of a beehive will start calling to a human, who responds if interested in a hunt. The bird continues to call as it moves through the forest and the person follows. Once the bird and the human are near the hive, the bird flies up in the direction of a hive living in a hollow spot inside a tree. The bird has done its share of the work, and now it's time for the person to do what the bird cannot do—access the hive to harvest the honey and honeycomb from it.

This is done by climbing the tree and using a tool such as an ax to break open the hive. The person is often stung while doing this, but the use of smoke to subdue the bees minimizes this risk considerably. Honey is such an energetically and nutritiously rich resource that it is worth some stings to acquire it.

Unanswered questions remain about how this association between birds and people evolved. Some people think that the bird's behavior evolved to allow a mutually beneficial relationship to exist between the birds and early humans. Others think that modern humans are taking advantage of an association that originated with birds guiding either baboons or honey badgers to beehives. Evidence of links with these species is minimal, and no complete guiding sequences with non-humans and the birds have ever been observed.

The bird has invested a considerable amount of time and energy into finding the beehive and leading the people to it. If the people don't repay the bird for its guiding services, the bird may refuse to guide that person in the future. Alternatively, according to legend, if you try to cheat the Greater Honeyguide, the next time it will lead you not to a beehive but to a lion's den.

Interactions between species are a great interest of mine. In fact, my first two scientific publications involved a nesting association between two species of tropical social wasps. I investigated the reason that they often lived just centimeters away from each other, and found that one of the wasp species was receiving protection from vertebrate predators from the other wasp species.

Creatures Great and Small

December 5, 2011

For a long time, paleontologists believed that animals typically evolve to become ever larger. Cope's rule, which states that animal lineages tend to increase in size over evolutionary time, is the best-known description of this propensity for increased size.

It's a lovely rule with one small drawback: it's not actually true. Many lineages do not show increased size over time. Recent research reveals that it is no more likely for there to be an increase in the size of animals in a lineage over time than it is for there to be a decrease.

Even with horses, whose average weight over the last 60 million years has increased tenfold and have long been the classic example illustrating Cope's rule, the apparent increase in size over evolutionary time is a result of considering only the surviving lineage of horses. Throughout much of its evolutionary history, the horse lineage contained species of many different sizes. However, all of the lines went extinct except for the modern branch of that tree, which contains the largest species. This is interesting because overall, larger species are more prone to extinction than smaller ones. In fact, one of the major advantages of small size is a tendency to be resistant to extinction.

There are other advantages of being small in size. Small animals have a better chance of surviving when resources are limited, they reproduce rapidly, and they can avoid natural enemies. Being too little to capture is a different strategy than being too big to take down, but it is effective.

Together, the many lineages with modern species significantly smaller than their ancestors demonstrate that Cope's rule is flawed. In numerous lines, the evolutionary trend was toward a decrease in size, and the results are some of the cutest animals ever.

Pygmy seahorses are less than an inch in length, and this is perhaps why the first species was not discovered until 1970 and new species have been identified as recently as 2009. Scientists think there are many more species of these tiny fish, which evolved from larger ancestors, but they remain undiscovered.

Among the little reptiles that evolved from larger ancestors are the Madagascar dwarf chameleon, which is less than an inch long; the Barbados threadsnake, which is 4 inches long and about as big around as spaghetti; and the speckled padloper, which is a tortoise that only reaches 4 inches.

Many lineages of mammals contain modern species far smaller than those from which they evolved. Pink fairy armadillos are less than 6 inches long and weigh about as much as a pear. At 4 to 5 inches tall, the Philippine tarsier is one of the smallest primates in the world.

All of these teeny tiny animals evolved from larger ancestors and are evidence that size can decrease in a lineage over evolutionary time. Cope's Rule suggests a consistent trend and direction in evolution that simply does not exist. The rejection of this idea illustrates a sentiment expressed by biologist T.H. Huxley: "The great tragedy of Science—the slaying of a beautiful hypothesis by an ugly fact."

The inaccurate ideas that (1) Bigger is necessarily better, and (2) Evolution represents some inevitable march of progress probably account for the human tendency to assume that animals become ever larger over evolutionary time. On another note, there is a movement to change the names of any species that currently include the term "pygmy" which is frequently used in a derogatory manner and is offensive in many contexts.

Don't Worry about the Reindeer
December 19, 2011

It's natural to worry about how reindeer are able to fly around the world pulling not only a sleigh, but a quantity of gifts that would weigh down a fleet of the largest Airbuses and a man who is a part of the worldwide obesity epidemic. Luckily, thanks to scientific research, we know a little about how reindeer survive the season.

One issue associated with hard work, even in winter weather, is the risk of overheating. Unlike toy-making elves, snowball-throwing children, and snow-shoveling adults, reindeer can't unzip their coats and let in the frosty air. However, scientists at the University of Tromsø (The Arctic University of Norway) have studied the ways that reindeer are able to cool down.

By training reindeer to run on treadmills in a range of temperature conditions, they were able to observe how these animals respond to excessive heat buildup. At first, they breathe heavily with their mouths closed to let cool air into their noses. This behavior cools the blood in the nasal passages, and the cooler blood moves around the body. When more heat loss is necessary, the reindeer open their mouths to pant, much as dogs do, which cools the blood even more and also allows for evaporative cooling. If their body temperatures are reaching dangerously high levels, reindeer are capable of moving the cool blood in their noses toward the brain to protect it from overheating.

The breeding season, which takes place in autumn, is very hard on male reindeer. They eat very little, and become quite thin, losing as much as a third of their body weight. It's worrisome to consider individuals in such poor condition making arduous, long-distance flights so soon after breeding. Once again, science comes to the rescue.

Male reindeer usually shed their antlers in the period from late November to early December. It's possible for a male to have antlers as late as December 24th, but it's unusual. From this, we can infer that Santa's reindeer are all female, since they don't drop their antlers until the spring, after giving birth. Another possibility is that they are castrated males (steers) since they also keep their antlers later in the season than breeding males do.

Of course, if breeding practices at the North Pole have resulted in the development of flying reindeer, then breeding for late retention of antlers hardly seems like an insurmountable obstacle. Flight has evolved at least four times: in birds, bats, insects, and the now-extinct pterosaurs. Unless reindeer are related to bats (and there is no evidence of a close connection between them), we have to consider the possibility that flight evolved a second time in mammals.

Though reindeer on the ground are sometimes killed by predators such as bears and wolves, there are no known flying predators of this species, so the worry that reindeer will be subject to attack during flight is unfounded. They are safer in the air than on the ground. Perhaps this is part of the reason they have thrived in the important seasonal role that has made them beloved around the world.

I'm told this column offered some relief to various people who were quite concerned about these working animals.

Unusual Mascots in Today's Rose Bowl

January 2, 2012

Lions and tigers and bears, oh my! These—along with cougars, bulldogs, eagles, huskies, panthers, and rams—are the most common school mascots. All are tough animals and to mess with them means risking considerable injury. It makes sense in the psychology of sports to have an intimidating mascot so that opponents inadvertently associate your team with aggression, force, strength, and stamina.

Of course, many schools have mascots that don't fit the usual mold. The mascot of the University of California, Santa Cruz, for example, is the not-so-mighty banana slug, and men's athletics at the University of Arkansas at Monticello are represented by the boll weevil, an insect measuring less than a quarter of an inch that is the most destructive pest of cotton. (Women athletes are known as the "Cotton Blossoms.")

Less extreme examples of unorthodox mascots are featured in today's Rose Bowl game between the University of Wisconsin and the University of Oregon. I attended high school in Oregon and got both my graduate degrees at the University of Wisconsin, so I know that people associated with these institutions take tremendous pride in their school mascots—the badger of Wisconsin and the duck of Oregon.

The Oregon Ducks were originally called the Webfoots, a nickname that had nothing to do with ducks. The "Webfoots" were a group of fishermen who were Revolutionary War heroes. Some of their descendants migrated to Oregon and the name came, too. As early as the 1890s, sports teams at the University of Oregon were known as the Webfoots, though it wasn't official until a 1926 naming contest. Live ducks were associated with the teams by the 1920s, and sports writers favored the term Ducks in their headlines. In the 1940s, the handshake deal between Walt Disney and Oregon's athletic director meant that Donald Duck's image became associated with the teams, and their mascot is officially known as "Donald Duck."

Usually, at least one team in any postseason football matchup will have an animal mascot that originates with the animal itself. Not so in this year's Rose Bowl. Just as the Ducks are historically linked to people, so are the Badgers. Wisconsin became known as the "Badger State" because lead miners in the 1820s and 1830s spent the harsh winters without proper shelters and had to "live like badgers" in tunnels built into the hillsides of their adopted state. In 1889, the badger became the official mascot of the University of Wisconsin, but his name was the result of a 1949 contest. Though the world knows him as Bucky Badger, his full name is "Buckingham U. Badger." For a few games, a live badger was in attendance, but it proved uncontrollable and fierce, so it was retired to the local zoo. At one point, a live raccoon with the name "Regdab" (badger spelled backwards) attended games.

Ducks and badgers don't commonly interact in the wild, so as a biologist, it's hard to predict a winner. On the football field, they are fierce and worthy opponents, so as a sports fan, I defer to the oft-quoted line, "That's why they play the game."

Animals are so commonly associated with sports teams that it is unusual when a school or team does not associate itself with some member of this kingdom. This speaks to the ways humans admire and connect with animals.

The Slow-Moving Sloth
January 16, 2012

It's winter and I feel slothful—tired, lazy, and desirous of sitting on my couch all day by the woodstove sipping hot tea. Actually, it's hardly fair to blame this on the season. My behavior is not entirely different from the slothful feeling I have in the summer when I want to lie in my hammock all day sipping ice-cold lemonade. One similarity is that both situations cause me to compare myself favorably to an actual sloth.

Sloths are aptly named. There is nothing vigorous about their behavior. Their metabolism is roughly half that of other similarly sized mammals. Doing everything slowly allows them to survive on their low-energy, low-protein diet, which consists of leaves, leaves, and more leaves. There are very few mammals that are able to survive on the poor quality of nutrition provided by an all-leaf diet, but the sloth has many adaptations that make this possible, including their slow-moving ways. (Sloths have a top speed of 0.15 to 1.2 miles per hour, depending on the species.)

Slowly and thoroughly chewing each mouthful of leaves allows them to extract more nutrients from the leaves. Their stomachs have bacteria in them that help with the digestion of the tough walls of plant cells. Like typical members of the primitive superorder Xenarthra to which they belong (and which includes anteaters and armadillos), they have simple teeth shaped like pegs. Those teeth are extremely strong with a blunt shape that allows them to mash up their never-ending salad of sustenance. They have exceptionally long intestines to maximize the nutrition they obtain from leaves.

Sloths hang out high in trees and only change to a new tree every day or two. They do this in order to vary the nutrients they take in. They come down from the trees even less often than they move among different trees. Approximately once a week, sloths descend close to the forest floor for the purpose of elimination. They dig a small depression with their stub of a tail, then urinate and defecate into it.

They are difficult to spot for many reasons. They move very little, and it's hard to see an animal in trees when the bright sky causes them to be in silhouette from many angles. They are camouflaged, thanks to the algae that grows on their fur and makes these forest animals appear to match the leaves around them. Many sources say that sloths move so slowly that moss grows on them, but it is algae, not moss, that gives their fur this greenish tint.

Sloths are good swimmers, so one way they move to new areas of the forest is by crossing rivers. Unfortunately, they are not nearly as capable when moving on the ground, so when they try to reach a patch of forest on another side of a road, many don't survive the crossing.

It's a tough PR problem when your species name contains one of the deadly sins, but it's hard to deny the appeal of an animal that makes my behavior in my most lazy moments seem positively energetic by comparison.

Sloth slowness has been made famous by the character Flash at the DMV in *Zootopia* whose lack of speed is highlighted to great comic effect.

Reptiles Join the Family

February 6, 2012

The general philosophy of life that prevails in our house is "the more the merrier." As such, our family expanded last month to include reptiles. We are now the proud owners of two leopard geckos.

Choosing a pair of herps (reptiles or amphibians) to be family pets was a challenging social exercise in consensus. One son really wanted a snake, and the other son wanted a frog, but in the same tank, the result might very well have been the following conversation between the kids:

"Where's Jumpy? I can't find him!"

"My snake looks swollen. Is she okay?"

No thanks!

We wanted pets that were relatively easy to care for, and who would do well in the 10-gallon terrarium that had been a Hanukkah present for our sons. It turned out that leopard geckos make the perfect pets for us. They are fairly small, easy to care for, inexpensive, and don't have high demands for attention. Though their natural habitat is the rocky dry regions of parts of Afghanistan, Pakistan, India, and Iran, they are commonly available at pet stores and do well as pets.

To start caring for our geckos responsibly, all we needed was a glass terrarium with a screen top, a lamp for heat during the day, a heating pad for inside the tank, accessories that provided places for them to hide, sand for the bottom of the tank, food and water bowls, and crickets and mealworms and a calcium supplement. They tend to defecate in the same spot, and cleaning the cage a couple of times a week and changing the substrate every month or so is all that's necessary in terms of normal maintenance.

The beautiful color patterns for which leopard geckos are named were a big factor in all of us being satisfied with the species that we chose. The scientific name, *Eublepharis macularius*, refers both to its color patterns ("macula" is Latin for spot or blemish) and to the fact that these animals are in the only sub-family of geckos with eyelids (In Greek, "eu" means true and "blephar" means eyelid.)

Though our pets came with a scientific name, choosing pet names for them proved to be far more contentious than selecting the species. The plan was to go with names that both children found agreeable, subject to the parental approval we were prepared to give for any names that were not offensive in some way. Relatively quickly, both kids accepted the name Rocky for one of them, but it took a long time to agree on a name for the other one, which meant that their conversation for several hours sounded like this:

Rocky and Granite? No!

Rocky and Tiger? No.

Rocky and Po? Hmm, umm, no.

Rocky and Peep? I like it, but not really.

Rocky and Quartz? Please, no.

Rocky and Lumpy? Maybe.

Rocky and Cinco? No!

Rocky and Bumpy? Silence. The kids considered these names and finally agreed on them both. Whew. If only members of Congress would follow their example and agree on something (anything!) as important.

Leopard geckos definitely make great pets, especially for people who don't yet have a lot of experience with reptiles but are interested in them.

Birds, Bees, Flowers, and Robbers

February 20, 2012

Many birds and insects are attracted to flowers because flowers produce nectar, an important food resource. Most visitors to flowers provide the crucial service of pollination, which results from their activities collecting nectar and pollen. Over the eons, both the flowering plants and their pollinators have influenced each other's evolutionary paths.

The changes in animal pollinators over evolutionary time have made them more adept at finding the flowers and more efficient at extracting resources when they are there. Flowering plants have evolved traits that better attract pollinators and that guide their foraging visitors to enter flowers. When they go inside flowers, they make contact with the floral parts in such a way that pollination occurs.

The co-evolution of pollinators and flowering plants is a tale of dependence and mutual benefits. They each offer something that the other requires, and they need each other. It's enough to make anyone feel amazed by the cooperation that exists between different species and the orderliness of the natural world.

Unfortunately, the idea that all of nature is in harmony is (how shall I say this?) a load of bunk. Among all this holding of hands and singing of Kumbaya are cheaters who exploit the system. Among those who take advantage of the pollination relationships are the nectar robbers. They visit flowers to feed on the rich resources there, but they don't pollinate the flowers. From the flowers' point of view, this is theft. For the animals that take resources without pollinating, it's simply foraging.

Making nectar is costly in terms of time, nutrients, and energy, all of which are in limited supply for the plant. The flowering plants produce these rich resources to attract the pollinators that are essential for their reproduction. The pollination services are worth the cost of nectar production to the plant. Nectar robbers take advantage of this system, exploiting it for their own benefit. They take the resources that flowers make to attract pollinators, but they don't provide a benefit to the plant.

There is more than one way to commit floral larceny. Primary nectar robbers make a hole in the tissue of the petals through which they remove nectar, and therefore bypass the routes taken by pollinators. Secondary robbers access nectar through holes made by primary robbers.

Nectar robbers may deter legitimate pollinators from visiting flowers because there is little or no nectar left, and they may further negatively impact plant reproduction by damaging reproductive parts of flowers.

Some nectar robbers always forage without pollinating, while other robbing species sometimes do enter the flower to forage and therefore act as pollinators. Animals that are nectar robbers, at least some of the time, include species among the ants, carpenter bees, bumble bees, pea weevils, stingless bees, flies, butterflies, hummingbirds, orioles, flower piercers, sunbirds, tyrant flycatchers, tanagers, and warblers.

The story of flowering plants and their diverse animal pollinators is a fascinating tale of cooperation, co-evolution, and mutual dependence. As in many good stories, there are villains, too, which in this case are the nectar robbers.

It is so important to understand that nature is filled with competitors. Even though there are so many examples of cooperation, that is just another strategy for competing successfully.

Thoughts on Animals
March 19, 2012

Politicians, authors, philosophers, comedians, and people in every other occupation all have something insightful to say about animals. It's no surprise, really, because animals have an impact on aspects of our life from the most serious to the most mundane, from the sublime to the ridiculous. It's natural for quotations and proverbs relating to animals to inspire as well as to entertain.

The greatness of a nation and its moral progress can be judged by the way its animals are treated.
—Mahatma Gandhi

If you think something small cannot make a difference, try going to sleep with a mosquito in the room.
—Unknown

Did you know that dolphins are so smart that within a few weeks of captivity they can train people to stand on the very edge of the pool and throw them fish?
—Unknown

The difference between friends and pets is that friends we allow into our company, pets we allow into our solitude.
—Robert Brault

A bird does not sing because it has an answer. It sings because it has a song.
—Chinese Proverb

Ever consider what pets must think of us? I mean, here we come back from a grocery store with the most amazing haul—chicken, pork, half a cow. They must think we're the greatest hunters on earth!
—Anne Tyler

An oppressive government is more to be feared than a tiger.
—Confucius

Man had always assumed that he was more intelligent than dolphins because he had achieved so much—the wheel, New York, wars, and so on—whilst all the dolphins had ever done was muck about in the water having a good time. But conversely, the dolphins had always believed that they were far more intelligent than man—for precisely the same reasons.
—Douglas Adams

Thousands of years ago, cats were worshipped as gods. Cats have never forgotten this.
—Unknown

The kind man feeds his beast before sitting down to dinner.
—Hebrew Proverb

I identify most strongly with the turtle: I patiently plod along till I reach my destination—and occasionally I stick out my neck.
—Paulette Peltan

We can judge the heart of a man by his treatment of animals.
—Immanuel Kant

Dogs feel very strongly that they should always go with you in the car, in case the need should arise for them to bark violently at nothing right in your ear.
—Dave Barry

My favorite animal is the mule. He has more horse sense than a horse. He knows when to stop eating—and he knows when to stop working.
—Harry S Truman

Lots of people talk to animals . . . Not very many listen, though . . . That's the problem.
—Benjamin Hoff

An ant on the move does more than a dozing ox.
—Lao Tzu

Our perfect companions never have fewer than four feet.
—Colette

Just because an animal is large, it doesn't mean he doesn't want kindness; however big Tigger seems to be, remember that he wants as much kindness as Roo.
—A. A. Milne

Animals charm us, as do the many wise words about them.

Animals make great muses, inspiring those with great minds to write with great wisdom.

Animals and Their Plants

April 2, 2012

Lately, my thoughts have turned to drugs. I've had few opportunities in my life to experience firsthand how incredibly effective pharmaceuticals are, but my recent oral surgery left me wildly impressed with them. The chemicals that humans have harnessed from nature to promote health and to ease pain as well as other symptoms of injury and disease are astounding.

People are not the only species on the planet using nature's pharmacy to achieve better living through chemistry. Observations of the plants that non-human animals utilize reveal a tendency of many species to make use of herbal medicines. The field of zoopharmacognosy concerns the uses of plants by animals to treat disease and the symptoms of it.

Medicinal plant use is best documented in primates. An entire new class of compounds, one of which has antibacterial, antiparasitic, and antitumor properties, was discovered because a field biologist observed chimpanzees in Tanzania consuming a plant containing it. Chimps only ate from this particular plant when they were ill, and analysis of their feces showed that levels of a parasitic worm were much lower about 24 hours after consuming its leaves. People living not far from these chimps also know to soak these leaves in water and then drink the bitter-tasting concoction when they have stomach pain, diarrhea, or other intestinal upset.

Woolly spider monkeys may be using plants in very specific ways to influence their reproduction. Two plants that they often eat are high in estrogen, so eating them may actually decrease fertility. In other words, they may be using these plants as a form of birth control. At other times, these monkeys have been observed eating monkey's ear, a plant which may increase a monkey's chance of becoming pregnant because it contains stigmasterol, a precursor to progesterone.

At the very end of pregnancy, elephants have been observed to walk more than five times their usual daily distance to find a tree in the Boraginaceae family. They then consume it in its entirety, perhaps to induce labor. Kenyan women seeking to induce labor use the leaves of this same tree to make a tea.

Many species engage in fur rubbing behavior in which parts of plants are applied directly to the fur or chewed and mixed with saliva and then ground into it. Though it occurs in bears, coatimundis, and spider monkeys, it has been best studied in capuchin monkeys. Capuchin monkeys use different types of plants, with citrus being the most popular. The plants they use are all strong smelling and most contain compounds with antiseptic, fungicidal, insecticidal, anesthetic, or anti-inflammatory properties. The plant most commonly used for fur rubbing by bears is called "bear medicine" and has long been used by people as an antibacterial agent and a topical anesthetic.

The study of people's traditional herbal medicines offers insights that lead to powerful pharmaceutical therapies. Studies of other animals and the ways they choose to medicate themselves may offer similar knowledge. It's exciting to contemplate how much humans have to learn about the use of medicinal plants from the many other species of animals that share the planet.

The use of plants as medicine is one more example of something that was once thought to be done only by people (such as tool use and language) but is now known to exist more broadly in the animal kingdom.

It's Alarming!

April 16, 2012

"Osprey!" I shrieked at our then 5-year-old son. He has always loved birds and I was eager for him to notice this majestic bird of prey that he had rarely seen outside of field guides. He came running out of the bushes faster than we'd ever seen him go. Regrettably, he had misinterpreted my urgency and thought I was warning him about the bird. Terrified, he raced toward us, wide-eyed and frantic, trying to avoid being captured by the talons of this predator by reaching the safety of his parents in time. I was just trying to let him know to look up so he didn't miss the opportunity to see a cool bird.

Though I was not giving one, alarm calls are common in social species, including monkeys, elephants, squirrels, meerkats, and blackbirds. Individuals acting as sentries or any other group member who spots a predator will warn others of the danger. Some scientists have hypothesized that alarm calls are altruistic because the individual who gives the alarm call may make itself more noticeable to the predator and become its target. In other words, the caller risks being attacked by raising the alarm that warns others. If this is true, it's hard to understand why the tendency to give alarm calls persists in so many species rather than dying out.

Perhaps the caller is attempting to save relatives, especially offspring, which makes the risk worthwhile. If this is the case, those who give alarm calls are helping to save other individuals who also have a tendency to give them, since they share so many genes. Support for this idea that alarm calls are preferentially used to save relatives comes from observations that in some species, individuals give alarm calls more often when surrounded by their own offspring or other close relatives.

Other data suggest that the caller is actually alerting other individuals to the predator so that more of them are aware of the situation and they can fight off the predator together. The alarm call may also communicate a message to the predator along the lines of, "Don't think you can catch me off guard, because I KNOW you're there."

Animals can convey specific information with their calls in addition to giving calls that basically mean, "Watch out!" Vervet monkeys have multiple alarm calls, and each call elicits a different response. A leopard sighting will make a monkey give a call that causes other members of the social group to run up into trees and onto the smaller branches that are too weak to support a leopard. The call given in response to an eagle makes nearby monkeys look up and take cover under the closest bush. Alarm calls given when a python is spotted have the effect of making monkeys stand up tall and look down.

If only I had thought to be as specific with my son. It would have been better if my call had clearly told him, "Amazing Osprey! Look up!" so he didn't think the message was, "Dangerous Osprey! Run for your life!"

The Osprey incident is a favorite family story. Thankfully, our son doesn't remember it, so he can enjoy the humor of it without reliving the fear he felt that day.

Wild Beasts, Children, and Art

May 7, 2012

When a critic referred to Henri Matisse's paintings as wild beasts, it was in no way meant as a positive remark. Room 7 at the 1905 prestigious Salon d'Automne became known as "the cage" and the artists whose works were displayed there were mocked. Their paintings were described as primitive, brutal, and violent. The bright, unblended colors of many paintings created in the early 1900s, with their bold strokes and lack of depth, were not well received initially. They are now recognized as belonging to the first avant-garde art movement of the 20th century and considered priceless treasures.

Referring to an adult as a wild beast or any other kind of animal will generally cause offense. Most grown-ups interpret such descriptions to mean uncouth, out of control, or even criminal. In social settings, being described as any type of animal is almost surely negative. Sometimes specific animals are mentioned, as in "He has all the grace of an elephant," or "She's like the proverbial bull in a china shop."

In athletic endeavors, the connotations are the opposite, so that to say, "She's an animal!" or "He's a beast!"

implies great skill, power, motivation, or drive. To say that someone moves like a gazelle indicates tremendous finesses and speed while having the "beauty and elegance of a swan" is high praise when referring to gymnasts, figure skaters, and divers.

Kids respond more consistently when faced with references to wild beasts, always assuming positive implications no matter what the context. They interpret being animal-like in a beautiful way, encompassing the very best of nature. During a Masterpiece Art lesson on Matisse in my son's first-grade class, the instructor, Jeni Jensen, started by asking the kids, "Has anyone ever called you a wild animal?" to which nearly all the students responded, happily sharing stories of when someone had referred to them in this way. They gleefully recounted details of the energetic antics and zaniness that had prompted some adult to compare them to a wild animal. They were proud of their energy, the joy, the activity, and the strength that prompted the comparison.

Children often feel great pride in sharing characteristics with any of the many wonderful animals that share our world. They are impressed by animals' strength, power, cleverness, dexterity, grace of movement, as well as their ability to fly or growl, leap, climb trees, or dig deep holes. Naturally, they consider it a compliment when any similarities between themselves and such talented animals are pointed out. Most children see animals as creatures worthy of envy and admiration, and therefore love to be associated with them.

Despite the negative intent, the "wild beast" comment launched Matisse's career. Eventually, he became known as a leader, along with fellow artist André Derain, of the Fauvist movement. The term originates from the French word *fauves,* meaning "wild beasts," and was actually coined by Louis Vauxcelles, one of the few critics who didn't pan Matisse's work completely. The critics and Matisse himself learned what most kids already know: it's cool to be a wild beast.

I love it when kids see the world of animals in such a good light!

Things Moms Say About Animals

May 21, 2012

Yes, Mother's Day 2012 is behind us, but I'm a mom the rest of the year, too, which gives me plenty of opportunities to say absurd things to my children. My kids, my friends, and I searched our memories for animal-related mom remarks and came up with these.

Your sand box is in the back yard. That's the cat's litter box.

What do you mean you found it in the dog's ear?

No, we are never going to have a pet elephant in this house.

Don't take the snake over to the neighbors unless they have asked to see it.

Sweety, we only ever flush the dead fish.

Hey, the gecko was not supposed to go in the soup.

Quickly boys, round up all of your Madagascar hissing cockroaches and put them back in their containers before our guests arrive.

We do not use the dog for T-ball practice.

I know the lake is bigger, but the fish are actually better off in our tank.

Don't put your watch on the dog. Trust me, he doesn't need to know what time it is as often as you do.

The guinea pig is not supposed to be roaming the house.

Yes, it's a remarkable skill and your reflexes are impressive, but I still don't want you catching birds from the feeder.

Please tell me that the reason the dog's mouth is red is that she ate your popsicle.

Why is there a frog in the oven?

Let's review! Crow outside, parrot inside.

Okay, who gave the dog a haircut?

We'll only be dying the eggs. The rabbit is staying her usual color.

Honey, we don't vacuum the cat.

Only eat the fish that we buy, not any from the tank.

I know YOU like scorpions, but we need to get him a birthday present that HE will like.

No fair blaming that gas on the dog. He's out on a walk with Daddy.

The mess in the playroom is so bad I thought the ferrets had escaped, but I checked, and they are all accounted for, which means you're in a heap of trouble and you have a big cleanup job ahead of you.

I just don't understand why you thought the horse needed my shoes. She has her own.

If you want to do turtle races, you'll have to start earlier next time because we are not delaying bedtime just so they can cross the finish line.

Where's your lizard? Why does the snake look so fat? Uh-oh!

The chickens will lay eggs when they are ready. Going into the chicken coop and pretending you're laying eggs will not speed things up.

When you tell people that we have a rat, please mention that it is a pet rat.

New rule: the hamster can never be a part of any arts and crafts project.

I don't care how many lives most cats seem to have. I don't want a single one of them lost today, so cut that out and get off the roof.

Have you talked to YOUR kids about animals today?

Moms tend to be a source of amusing things to say in many areas of life, and their statements about animals are no exception.

Domestication and the Anna Karenina Principle

June 4, 2012

There have been many attempts to domesticate the zebra, none of them successful. Though this has surprised each group of people who have tried, the failures are predictable.

The potential for domestication follows the Anna Karenina principle, which describes an undertaking in which a lack of any one key component results in failure. A successful undertaking subject to this principle is one in which every such deficiency has been avoided. The name refers to the opening line of Tolstoy's novel *Anna Karenina*: "Happy families are all alike; every unhappy family is unhappy in its own way."

An animal species must have certain characteristics to be domesticated, and the lack of any one of them is a deal breaker. Very few species have all of the necessary traits. They need to be easy to feed rather than be picky eaters. They have to grow quickly and be able to reproduce in captivity. They can't be prone to panic or to fleeing in response to fear. They cannot be ill-tempered or unpredictably violent. They must have a flexible social hierarchy.

Zebras have a tendency to attack people and to become more aggressive as they age, which has made their domestication impossible. They have never lost the nasty part of their disposition that includes unexpected biting. It's not just zebras on the list of domestication failures. African buffalo and grizzly bears also have unpredictable temperaments that preclude their domestication. Deer are very flighty. Their tendency to bolt in a panic is incompatible with domestication.

There are members of all of these species of "domestication failures" that are calm and friendly with humans, but such animals have been tamed, not domesticated. Despite the fact that these terms are sometimes loosely used to mean the same thing, taming and domestication are not synonymous. They are, in fact, very different processes.

A wild animal that has become accustomed to humans and human contact, and is not fearful or aggressive around them is a tame animal. Taming is the process by which an individual animal becomes used to people and tolerates them being close and even having physical contact with them. There are no genetic changes associated with taming, and it is only relevant within the lifetime of the individual animal affected.

In contrast, domestication is an evolutionary process involving an entire population of animals that takes place over many generations. Species that have undergone domestication are genetically different than their ancestors. Those genetic changes affect morphology, physiology, development, and behavior.

An individual animal (perhaps a raccoon, monkey, peccary, or fox) that has been taken in by humans and has learned to be comfortable as well as gentle around people, is tamed. In contrast, a species of animals (say cows, pigs, dogs, or horses) that has changed genetically—with corresponding morphological, physiological, developmental, and behavioral changes that enable it to live with humans and have its breeding controlled by humans—is domesticated.

Far more species of animals are amenable to taming than to domestication. The Anna Karenina principle provides a compelling reason why this is so.

The Anna Karenina principle was made popular in biological circles in Jared Diamond's 1997 book *Guns, Germs, and Steel: The Fates of Human Societies.*

Unexpected Insect Outbreaks
June 18, 2012

Insect outbreaks often appear unexpectedly, whether it's bark beetles killing acres upon acres of pine trees, ladybird beetles invading our houses, armyworms stripping wheat fields bare, or gypsy moth caterpillars turning oak leaves into a steady rain of caterpillar droppings. The obvious question is "Why do insect outbreaks occur?"

Several insects are known for their abrupt increase in abundances. In the eastern United States, the gypsy moth has a habit of appearing suddenly within a forest before disappearing again for many years. This insect voraciously feeds on the leaves of trees and shrubs, threatening residential trees, forests, and agricultural tree crops. Gypsy moths have garnered more notice, frustration, research dollars, and environmental concerns than any other forest or urban tree pest in the eastern United States. Most of the time, natural controls keep gypsy moth populations in check. These include predators, parasites, disease organisms, and even bad weather, all of which tend to maintain low populations of the moth.

However, if something happens to alter this balance and the survival rate of the insect increases, it can trigger a sudden increase in their population. Because one female moth can lay up to 1,000 eggs, gypsy moths are capable of rapid population increases. All it takes to maintain the normal population level is for one male and one female out of that 1,000 to survive. That's only a 0.2 percent survival rate. To put it another way, 99.8 percent of the eggs and larvae can die without reducing the total population.

In northern Arizona, we have a native insect called the Pandora moth that also appears unexpectedly. Every 20 to 30 years, large populations of this insect overtake the ponderosa pine forests north of the Grand Canyon. Interestingly, this insect has a 2-year life cycle with adult moths occurring during even years (2012, 2014, etc.) and caterpillars during odd years (2011, 2013, etc.) Because defoliation happens only every other year and trees are able to recover in alternate years, the damage caused by caterpillars does not kill pines. However, the defoliation can be unsightly and diminish the experience of visitors to the canyon.

Pandora moths started to outbreak near Jacob Lake last summer, and many caterpillars were seen feeding on ponderosa pine needles. This summer, expect to see thousands of moths near Jacob Lake in mid-August, as these caterpillars metamorphose into adults. Why there are periodic outbreaks of this insect is not known, but a virus is believed to cause populations to decline. The Paiute people in California used to celebrate these outbreaks and would harvest, prepare, and store the larvae (which they call piuga) as a preferred food.

Insect outbreaks appear to happen overnight because we generally don't notice the increasing numbers until large areas are affected. Fortunately, natural biological controls or harsh weather conditions can reduce insect abundances. Eventually the booming population crashes, and the insects that were so numerous seem to disappear as mysteriously as they arrived.

This is the only column I co-wrote, and the other author is my husband. It was also the only time the byline was so detailed about my life. Here's what appeared in the paper the day this column did:

Richard Hofstetter, PhD is an Associate Professor of Forest Entomology in the School of Forestry at NAU.

Karen London, PhD is an Adjunct Faculty in the Department of Biological Sciences at NAU. Rich and Karen co-teach a field course in Nicaragua (Tropical Forest Insect Ecology) and a freshman seminar (Sex,

Bugs, and Rock 'n' Roll). They took an ethology class together (1991), officially met at a seminar on the biology of sweat bees (1993), started dating (1994), married each other (1999) and have had two sons (2003 and 2005), all since the previous Pandora moth outbreak in the 1980s.

Additional notes: The insect that causes so much damage in our eastern forests has been renamed the spongy moth—an improvement as the new name is free of a racial slur. Rich has since been promoted from Associate Professor to Professor.

Food Caching and Retrieval
July 2, 2012

As we exited the mall one day when my kids were just 3 and 4, I said to them, "Do either of you remember where we parked the car?" A man who overheard me complimented me for providing this educational opportunity to my children, but after he was out of earshot, I said hopefully, "Okay, seriously, do either of you remember where we parked the car?"

Most humans have similar lapses in memory, which is why we're always searching for our keys, desperately looking for our cell phone, or even misplacing the important paper that was in our hand just minutes ago. We're very different in this way than animals that "cache" food items in multiple places and retrieve them much later, because their spatial memory is incredible. Cached food is only valuable if it can be found and eaten, and cachers are capable of finding thousands of individual food items months after burying them.

Caching food is a way to handle seasonal variation in food availability. For example, animals that cache acorns do so when there are too many to eat all at once and retrieve them when the trees are done producing them for the year and food has become scarce. Since seeds maintain their food quality for a long time after burial, seed-eaters are often cachers. Proficient cachers include chickadees, jays, ants, nutcrackers, and chipmunks. "To squirrel away" means to save up for future use, and squirrels are perhaps the best known of the animals who exhibit this behavior.

It's common for animals to watch other individuals, including squirrels, caching food in order to pilfer those resources later. The theft of food items can lead to big losses in an animal's food supply. Therefore, it is hardly surprising that in many cases, if a squirrel notices that he is being watched while hiding food, he will come back later to retrieve that food item and re-cache it elsewhere.

Reburying a food item is not the only cunning trick that squirrels employ to prevent having their hard work benefit someone else rather than themselves. If they have lost a lot of nuts or seeds to thieves who have taken them after burial, squirrels become even sneakier. They will pretend to bury a nut in a hole but not really do so. They may dig many holes before burying anything in one. Or they might try to hide food in a place where it's more difficult for anyone to watch them, such as under a bush or in a tree.

This kind of deceptive behavior is just one way that squirrels demonstrate their mental capabilities. Additionally, they are more likely to bury acorns from red oaks, which tend to last a long time, but they often immediately consume the more perishable white oak acorns.

If you remain unconvinced and are looking for more proof of squirrel cleverness, consider this. Have you ever, even once, seen a squirrel wandering aimlessly through a giant parking lot looking for a misplaced car? I didn't think so.

My sense of direction remains epically terrible, and I still rely on my kids to point me in the right direction in parking lots, airports, on the road, and everywhere.

Birding by Numbers

July 16, 2012

In my husband's family, birding is serious business, and I fit right in. No holiday passes without bird feeders or birdseed gifts, and bird clocks, bird quilts, bird socks, and bird books have all made appearances. Every one of us has, at least once, walked off the porch in pajamas and slippers to follow an elusive bird and ended up an embarrassingly long way from home considering our attire.

Birders' great moments usually involve sightings of extraordinarily rare or beautiful birds—the Resplendent Quetzal, the Vermillion Flycatcher, the Hyacinth Macaw, the Double-Wattled Cassowary, the Rainbow Lorikeet, the Harpy Eagle, the Elegant Sunbird, the Painted Bunting, and the Banded Cotinga appear on many lists of favorites. Many who love birds also have a favorite common backyard bird such as the Mountain Chickadee, the American Goldfinch, the Tufted Titmouse, or the Northern Cardinal.

For many birders, it's not just the quality of the birds spotted, but also the quantity. In a bizarre competition, birders doing a "big year" try to see as many species of birds as possible in a single calendar year within a set geographic area. The record for the American Birding Association area is 745 species, which was set in 1998 by Sandy Komito. The area involved is the 49 continental U.S. states, Canada, 200 miles of bordering ocean and a complicated set of rules for nearby islands that excludes Hawaii, Bermuda, the Bahamas, and Greenland, but includes the French islands of St. Pierre and Miquelon.

A valiant attempt to break this record was made by John Vanderpoel in 2011, but he fell just short with 744 birds. The total number of species that could possibly be seen in the area of the competition, including regular visitors, breeding species, accidentals, casual visitors, and exotic species that are now established here is currently 970.

Our family has attempted a Flagstaff "big day" on several occasions. A big year had its appeal, but that kind of financial and time commitment was too much for us to take on. People who do a big year typically fly close to 100,000 miles and drive around 75,000 miles, and I don't even want to think about how much money they spend. Our best big day this past winter involved sighting 33 bird species, which is pretty small compared to the record of 264. That was for a conservation fundraiser that brought in around $200,000 and was made possible by weeks of scouting and preparation by a large team of people to create a streamlined, efficient route from one hotspot to another while birding from midnight to midnight in Texas on April 22, 2011.

Perhaps more big days or even big years are in our kids' future. They certainly got an early start at birding. When our oldest son was born, my husband called his brother for parenting information. You might wonder whether his first question involved health, nutrition, safety, or some other common concern of new dads. No, his first question was, "How old were your girls when they could start using binoculars?"

With older kids and more knowledge of birding hot spots, our family's more recent big days have been significantly better—with reports of 128 species in a single day around Coconino County.

Citius, Altius, Fortius

August 6, 2012

The Olympic motto—"Citius, Altius, Fortius"—means "swifter, higher, stronger" when translated from Latin and embodies the goals of the world's best athletes. Of course, I am being very species-centric when I say "the world's best athletes." Although humans are quite extraordinary in what they can accomplish, members of other species perform some of the most astounding physical feats.

Swifter

Usain Bolt may move so fast he appears to have an engine, but his top speed of nearly 28 miles per hour isn't going to impress the average cheetah. These predatory cats can sprint at close to 70 miles per hour, which has earned them the distinction of the fastest land animal. The fastest animal in the water is the sailfish, which has been clocked swimming at 68 miles per hour. Missy Franklin is young and likely to get even faster, but at her best, she swims in the neighborhood of 4 miles per hour. The Peregrine Falcon is truly the fastest animal on earth, able to fly at speeds over 200 miles per hour in a stoop (dive). For comparison, divers in the 10-meter platform competition hit the water at around 35 miles per hour.

Higher

McKayla Maroney gets 2 extra feet of height on her vaults compared to everyone else competing, and it's impressive to see her soar up to scary heights. ("Scary" is a technical term in MomSpeak.) Pole vaulters routinely propel themselves to incredible heights, with the record being a shade over 20 feet. On the other hand, fleas can jump 8 inches high (without the help of a pole), which is equivalent to Superman's feat of jumping over tall buildings. The record for "higher" has to go to the Bar-headed Goose, which flies rather than jumps toward the sky. It breeds in Central Asia and winters in South Asia. Its migration path takes it from sea level to more than 27,000 feet as it flies over the Himalayas. Scientists are unsure why these geese fly so high instead of going over lower Himalayan passes as other migrating bird species do.

Stronger
Om Yun Chol, who weighs 123 pounds, won a gold medal last week lifting 370 pounds, which means that he was able to hoist three times his body weight up from the floor and then over his head. This compares favorably with the grizzly bear, which can lift 1,200 pounds, or 80 percent of its 1,500-pound body weight. However, this champion performance does not even touch the capabilities of rhinoceros beetles. These large insects can lift 850 times their own weight, which is perhaps the reason that beetles in this subfamily are also called Atlas beetles or Hercules beetles.

Great performers are not limited to a single species on this planet. Olympians push their bodies to excel in astonishing ways day in and day out for years, but we tend to marvel at what they do only during the Olympics. Perhaps it would be wise to remember to be amazed every day of every year by what each species can do, including our own.

I was happy to share my name with the Olympic Games in 2012 and liked to pretend that "London 2012" had something to do with me.

Swimming with a Whale Shark
August 20, 2012

One of my annual summer goals is to have an experience worthy of the dreaded back-to-school essay, "What I Did During My Summer Vacation," even though I'm no longer officially required to complete such an assignment. This summer, I succeeded by swimming with a whale shark. The name "whale shark" reflects commonalities with many whales, such as being enormous in size and filter feeding, but it is a shark and not a whale.

People travel from all over the world to the few places where whale sharks are reliably spotted for the thrill of swimming with one. It's the largest fish alive today, reaching lengths of more than 40 feet, but despite their gargantuan size, these gentle animals pose little threat to humans. They eat mostly plankton, although they also consume some small fish, squid, and crustaceans. They take huge amounts of water into their mouths, which they then close, forcing the water out through their gills. The structure of their mouths acts like a sieve, straining the water and trapping even tiny organisms.

Their quest for food makes whale sharks easier to find than many other marine species. They live primarily in the open ocean, but migrate seasonally to feed in shallow areas with abundant food. They are commonly seen on the coast of the Mexican state of Quintana Roo, and that's where I swam with one.

We were on the boat when we first spotted the dorsal fin breaking the surface of the water, which was exciting enough. Being in the water with the animal was even more thrilling. The visibility underwater was so poor that once I was close enough to see it, I was only 6 or 7 feet away, which is the perfect distance for truly appreciating the whale shark's size and beauty.

The mouth was 4 feet across and the five gill slits on each side were so extraordinarily large that I had to fight the irrational fear that I would be sucked right into them. The tail, which can injure humans who are hit by it, was approximately my size. The ridges along the sides of its body as well as the golden spotted and striped pattern on the gray-blue skin were beautiful. These patterns are unique to every individual and used by scientists for identification. The wonder of swimming alongside a whale shark was unforgettable. There were remoras and hundreds of other fish swimming with the whale shark, too, so I felt as though I were part of an oceanic parade.

This kind of experience could lead many people to have profound thoughts about our place in the universe or to create beautiful poetry on the spot. I wish I could say that my articulate side was stimulated, but it wasn't. What went through my mind was, "Wow!" which was quickly followed by "Whoa!" and then "Wow" again, followed by yet another "Whoa!"

Despite Yoda's assertion that "Size matters not," after swimming with a whale shark in the open ocean, it's inescapable that size (its most memorable feature) does matter.

This experience still ranks near the top of my "best ever" wildlife encounters. I was closer than was wise or fair to the shark, but it was an accident due to the poor visibility. Please forgive me.

They Look Alike, But Why?
September 3, 2012

Famous people who look alike may be related, but not necessarily. The Baldwin brothers Alec, Billy, Stephen, and Daniel are family, but Tina Fey and Sarah Palin are not. Likewise, similarities between species may or may not reflect a common ancestry.

An example of similarities resulting from a shared ancestry is the bone structure in the limbs of vertebrates from groups as diverse as frogs, lizards, turtles, birds, primates, rodents, seals, cats, and elephants. They all have two pairs of limbs, each with one bone in the upper limb and two in the lower limb and five digits. Many adult animals do not have five digits, but they still develop from an embryonic stage that does. Despite the huge differences between the limbs of horses, sea lions, monkeys, and bats, these diverse animals share the same basic structure. The divergence of those traits in different species is the result of evolution from a shared ancestor, and they are called "homologous" structures.

Sometimes traits in different animals appear so similar that it's natural to assume that they are homologous, but in fact they are not. Some similarities are a result of a similar function rather than a shared evolutionary history. Such structures are called "analogous" and arise by convergent evolution, which is the process by which different lineages evolve a similar characteristic or appearance without being related to one another.

For example, bird wings are analogous to the wings of insects. The wings are similar because they have the same function. To fly, an animal needs to be able to produce lift, which can be accomplished by having large thin structures extending on both sides. That's why all animal wings, as well as the wings of planes, are shaped the way that they are.

Similarly, the shape of dolphins, fish, and the extinct ichthyosaurs resulted from convergent evolution. Aquatic animals need to be able to move through the water efficiently, which is best done with a streamlined shape.

There are many structures that appear not to have a function, and these are referred to as vestigial structures. Many vestigial structures are homologous to functional organs in related species. Both snakes and whales have remains of the pelvis and hind limbs. Flightless birds typically have wing remnants. Humans have a coccyx, which comprises several fused tail bones although the entire structure is contained inside the body. Our coccyx is homologous to the tails of many of our primate relatives.

The human appendix was long considered homologous to the caecum of many of our mammalian relatives and without function (unless you count creating opportunities for a lot of pain and emergency surgery as functions). We now know that it plays a role in immune function and that it has evolved more than once. The appendix in humans, other primates, and rodents is actually analogous to the appendix in marsupials.

Homologous structures are like the faces of the Baldwin brothers—similar because of relatedness. Analogous structures and convergent evolution are more accurately compared to the famous likeness between Sarah Palin and Tina Fey.

I wish I had included in this column that bat wings are analogous to bird wings and to insect wings because I received a number of questions asking me about the relationship between bat and bird wings.

Endemic, Native, Introduced, and Invasive Species
October 1, 2012

If you want to see lemurs in their natural habitat, the only place to do that is on the island of Madagascar. In scientific terms, we say that lemurs are endemic to Madagascar, which means that they only live on that island. Islands typically have lots of endemic species because those species have evolved in isolation over many millions of years. In addition to its famous lemurs, Madagascar has more than 100 species of endemic birds and nearly 300 species of endemic frogs.

Australia also has large numbers of endemic species, which is predictable based on the isolation of this continent. It separated from Antarctica roughly 40 million years ago and has been on its own ever since. Animals that are endemic to Australia include koalas, kangaroos, numbats, emus, platypuses, bandicoots, Royal Penguins, wallabies, echidnas, Budgerigars, wombats, and a number of lovely fairy-wrens, including the species whose actual name is the Lovely Fairy-wren.

The Galápagos Islands are famous for their endemic species, which include the marine iguana, the giant tortoise, and the Flightless Cormorant. People come from all over the world to see these and other rare endemic species.

High rates of endemism are typical of geographic regions, such as remote islands, that are biologically isolated. Though such islands typically have a high proportion of endemic species, they also have species whose distribution extends beyond their shores. The species that are present in a region as a result of natural processes rather than as the result of human intervention are called "native" or "indigenous" species. All species have a natural range and within it, they are considered native.

In Hawaii, for example, the naturally occurring Ruddy Turnstone, Short-eared Owl, and Pacific Golden Plover are all species whose members live in other places, too. Hawaii, as expected for a group of remote islands, also has many endemic species. Roughly a quarter of all Hawaiian fishes and birds are endemic. Endemic species are at far greater risk of extinction than native species, which have larger distributions.

If people have brought species to areas outside of their natural range, they are referred to as introduced species in those areas. Sometimes introduced species have disastrous effects on the native species. For example, the brown tree snake was introduced to the island of Guam, most likely as a stowaway in military cargo in the years after World War II. By the 1960s, the snake had decimated many of the native and endemic bird species, which had evolved in the absence of snake predators and had no defenses against them. Populations of bats suffered similar losses—the last sighting of the Marianas fruit bat on Guam was in 1968.

The brown tree snake is an invasive species, which means that it is an introduced species that adversely affects the ecological or economic health of its new habitat.

In addition to direct economic damage, invasive species around the world have caused the extinction of thousands of species of animals, many of which died out without us having the opportunity to learn very much about them at all.

Shout out to an endemic animal (and a family favorite) near my home in northern Arizona: The Kaibab squirrel is endemic to the ponderosa pine forests on the north rim of Grand Canyon National Park and northern section of the nearby Kaibab National Forest.

Echolocation by Bats

October 15, 2012

Bats have long been associated with Halloween, largely because they look odd, drape their wings around themselves like a witch's cloak when they sleep, and have the spooky habit of being active at night. Hunting after dark means that insect-eating bats can feed on a plentiful food source with less competition than at other times of day while being exposed to fewer predators themselves.

The challenges of hunting at night include compensating for the inability to see well. Bats have solved this problem by using sound to navigate and find prey.

It has been known since the 1790s that bats use sound to find their way. The Italian scientist Lazzaro Spallanzani conducted experiments in which he tampered with bats' ability to see, smell, or hear. Bats were unimpaired when their vision and olfactory abilities were blocked, but if he plugged one or both ears, the bats were unable to navigate and avoid objects.

Plugged ears interfere with bats' navigational abilities because they are using echolocation, which uses the same physical principles as sonar. Bats emit sounds in the ultrasonic range, which are higher in frequency than humans can perceive, and listen to the returning echoes. The sound waves hit objects in the environment and bounce back, but the size, shape, motion, and material of the objects they hit change them in characteristic ways.

Though people can generally only get very basic information from echoes (I'm in a narrow canyon, I'm in a wide canyon, I'm not in a canyon), bats are able to perceive tremendous details of their environment from the echoes. They can navigate through narrow spaces and around small obstacles, and determine the type of prey that is flying around as well as its size and direction of flight.

Though people had been aware for a long time that bats use sound to navigate, it wasn't until the 1930s that American scientist Donald Griffin reported three distinct types of evidence that demonstrated definitively that bats use ultrasonic calls to navigate.

One, he used equipment, new at the time, that converted ultrasonic calls into frequencies that humans can hear, and found that bats do emit high-frequency calls, and also that they increase their calling rate as they approach obstacles. Two, he demonstrated that bats are capable of hearing sounds up to 98 kilohertz, which was as high as the equipment could measure at that time. (Bats actually hear up to about 150 kilohertz, compared to around 20 kilohertz for humans.) Three, in a series of experiments involving bats flying in a room with wires suspended from floor to ceiling, he showed that bats must be able to both produce sounds and hear them in order to navigate around the wires. Bats whose abilities to make or hear sounds were temporarily obstructed navigated poorly through the wires, but bats whose eyes were covered suffered no decline in performance.

What bats are capable of doing—foraging successfully at night, even in total darkness—*is* a bit spooky, but in an amazing gee-whiz science way, not the typical Halloween way.

Sadly, bats are much maligned and generally unpopular, but most people develop a more positive view if they understand echolocation and if they are told that many species of bats eat massive quantities of mosquitoes.

Getting Skunked

November 5, 2012

One of the few things that my husband and I both remember from organic chemistry is that the compound that makes skunk spray smell the way it does is butane thiol. We're both biologists and the aspects of chemistry that relate to ecology, evolution, and behavior stick with us most, even 25 years after our chemistry courses.

The smell of skunk also sticks with us, as we breathe it in almost every night around our house. We are not alone. You're bound to catch a whiff of skunk in just about any neighborhood in Flagstaff, and if you have a dog, there's a good chance your best friend is going to be sprayed eventually. You can hardly say "skunk" in Flagstaff without everyone in spitting distance sharing the story of an encounter with one. Many tales involve a skunk spraying through the screen door and into the house.

Supposedly, they spray only as a defense, but I'm not aware of any evidence that it's the smell of skunks that prevents screen doors from preying on these animals. (On the other hand, I know of no cases of skunks being attacked by screen doors, so maybe it is working as a deterrent.) Skunks can spray accurately to a distance of 15 feet, and it is a defense mechanism that is good enough to deter wolves and foxes, who generally avoid preying on skunks because of their foul odor. Owls are the main predators of skunks, and like most birds, they have a poor sense of smell.

That horrible smell, that butane thiol-y aroma, comes from oil glands under skunks' tails. Skunks can only spray five to eight times before running out. It can take a week to rebuild their supply if depleted, so if several of your neighbors have been hit by the same skunk recently, perhaps you'll be safe for a brief time as long as there are not other skunks whose beat is your block.

If you see a skunk who is stamping its feet, hissing, growling, or baring its teeth, you should assume that the skunk is warning you to back off and leave it alone. Those actions are the behaviors that skunks do before spraying. If they remain agitated and feel the need to act defensively, those warning signs may precede the handstand with the back end up in the air, the bending of the back into a U shape, and spraying. If you back up and leave it alone, you increase your chances of escaping unscathed.

Because of their smell, skunks are not universally appreciated for what they do ecologically. They play a large role in controlling both insects and rodents. They are omnivores whose diet includes these items as well as eggs, plants, fruit, carrion, and reptiles. They even successfully attack beehives to feed on the bees and the honey.

The most common species of skunk in Flagstaff is the striped skunk. The spotted skunk and the hog-nosed skunk are also seen in this area, but far less frequently. Perhaps those species fear our screen doors.

I absolutely love seeing skunks in our neighborhood, whether they are at the pond in our front yard, in the garden, or near a neighbor's house. The joy of a sighting more than makes up for the smell of these animals.

Camouflage and Hide-and-Seek
November 19, 2012

Caterpillars exist in a life-or-death game of hide-and-seek. These creatures are considered the most succulent, delicious meals that nature has to offer by birds, rodents, lizards, spiders, beetles, and a variety of other predators. Caterpillars, quite rightly, act to survive in order to turn into butterflies or moths, a fate that is obviously preferable to becoming food for other animals.

Caterpillars are abundant as well as slow moving, yet most of them go undetected each day, remaining safe from predators. Many are difficult to spot even at close range due to their effective camouflage. Camouflage is visible deception that makes prey hard for predators to detect, and caterpillars make use of it in a variety of ways.

Many of them employ camouflage to hide by resembling their environment. This makes them blend into the background so predators are unable to find them. A lot of caterpillars are the color of the substrate on which they live—green if they are on leaves, or brown if they are on bark, for example. No matter what substrate these caterpillars are on, their color patterns are usually mottled so that their outline appears broken up, which is an important aspect of blending in. (The same principle applies to military clothing and equipment.)

Rather than hide, some caterpillars are clearly visible but avoid being eaten by looking like something unpalatable instead of looking like dinner. The variety of objects they resemble is great, and the degree to which many look exactly like something else is striking.

There are caterpillars that look like leaves, mimicking them in shape and color pattern, complete with faux leaf veins. Even more specific are the caterpillars that look like a brown spot on a leaf. Wherever they are on a plant, they look more like a dead part of the foliage than a tasty caterpillar. Others are very hard to distinguish from real twigs. There are caterpillars that look like snakes, with a back end shaped like a snake's head, including eyespots resembling a snake's eyes. Being mistaken for a snake means that many caterpillars are avoided altogether by potential predators.

Among the most unusual features of the environment that caterpillars mimic are bird droppings. Such caterpillars are dark brown with mottled creamy and yellow markings and a lumpy, irregular shape. They are easily mistaken for bird poop even *after* you know that they are caterpillars.

Members of one species of caterpillar create their own disguises rather than being naturally camouflaged. They pick flowers from the plant on which they are feeding and attach them to their own bodies with silk.

They periodically replace petals that have gotten old and started to wilt.

Whether caterpillars use camouflage to blend in with the background or to resemble something that is not appetizing, their ability to look like something other than a caterpillar provides an excellent form of protection. Though caterpillars are not the only animals to use camouflage, they are among the animal kingdom's true masters of disguise.

I love finding camouflaged caterpillars, and I love it even more when someone points one out that I failed to find. Their ability to hide in plain sight evolved to fool all kinds of animals, including vertebrates like us.

Schooled by Fish
December 4, 2012

We had become one with a school of fish. Thousands of them swam with us so that we were surrounded. They were in front of us, behind us, under and over us, and on both sides of us. The number of fish was extraordinary.

The healthy coral reef of Gili Trawangan, a small island between Lombok and Bali in Indonesia where my husband and I traveled last month, supports a fish population of unimaginable size, and many of them are in schools. For the sake of fun with jargon, fish that are together for social reasons of any sort are "shoaling" and shoaling fish that are also swimming together in the same direction in a coordinated way are "schooling."

Fish in groups benefit because there truly is safety in numbers. Being in a big group means that there is not much chance of any individual fish being the target of a predator. Additionally, being in a large group of fish swimming close together presents challenges for a predator attempting to focus on a single individual. The large number of tails, fins, heads, eyes, and bodies moving quickly through the water make it hard for a predator using vision to zero in on a single individual in order to capture it.

Besides the anti-predatory benefits, schooling is thought to increase swimming efficiency. By drafting off the currents of fish in front of them, fish in schools need less energy to travel the same distance as a fish swimming alone.

The benefits of schooling behavior are easy to understand. It's far more difficult to comprehend how fish are able to move in such a coordinated fashion without the assistance of either a choreographer or a director. The school changes shape continuously, morphing in a manner that would challenge the best marching band, but seems effortless to the fish. The hundreds or thousands of fish in a school repeatedly change direction in a synchronized way, all seeming to respond simultaneously to some external cue. Actually, it's the actions of individual fish making split-second decisions based on their own perceptions of their nearest neighbors in the school that create these coordinated movements of the whole group.

Mechanoreception (the perception of mechanical stimuli that trigger the senses of touch, balance, and hearing) is critical for maintaining the integrity of schools. The lateral lines of fish are very sensitive to changes in vibrations and currents, some of which result from the swimming motions of nearby fish. Lateral lines extend along the length of the body of many fish, provide information by sensing changes in pressure and movement of the water, and are essential for maintaining schooling behavior.

I've often imagined how glorious it would be to be part of the ocean's world, and this is about as close as I've gotten. It was great being among the fish, seeing them change direction and speed together, and avoiding collisions with other fish that were so close that it's hard to see between them, even when I was (temporarily) part of their school.

I have always felt at home in the ocean and am happiest when I am there. Being surrounded by thousands of fish while snorkeling in Indonesia made me feel completely at home in a way that perhaps only other people who love the ocean as much as I do can relate to.

Out of Many, None

December 18, 2012

Last week, I found myself in the middle of a big crowd of holiday shoppers all heading for the same bargains at a sale. Though my first thought was that we resembled dairy cows heading to the barn, I resisted the urge to moo, fearing nobody would think it was funny. I'm grateful for this self-control, because a few minutes later, I concluded that the herd of cattle analogy was inadequate.

After much consideration, I decided the group of animals that best compared to the absolutely enormous numbers of people surrounding me was a giant flock of Passenger Pigeons. These birds were so abundant that they used to darken the skies as they flew over in massive groups of birds for days at a time, but now cannot be found, much like crowds of holiday shoppers after the first of the year. Sadly, instead of just dropping in abundance seasonally as shoppers do, the Passenger Pigeon is extinct, which is truly remarkable considering how many of them there once were.

One flock of Passenger Pigeons observed in 1866 in southern Ontario was estimated to have been a mile wide and hundreds of miles long. It took over half a day to pass overhead and may have contained 3.5 billion (yes, billion with a "b") birds. It is generally thought that Passenger Pigeons made up 25–40 percent of all birds in the United States.

Nesting areas could cover hundreds of square miles with hundreds of nests in a tree. A single egg was laid into nests loosely constructed of twigs, and both parents incubated the egg and fed the chick after hatching.

The Passenger Pigeon most closely resembled the mourning dove in size and appearance. It was slate gray-blue on its head, back, and wings, with a red breast and bright scarlet eyes. It was about 16 inches high.

The scientific name of this species is *Ectopistes migratorius*. Ectopistes means moving about or wandering and migratorius means migrating. The full name refers to the Passenger Pigeon's habit of not only migrating in spring and fall, but also of moving to choose different areas each year based on the best spots for feeding and breeding. Passenger Pigeons wintered throughout the southeastern United States, while their nesting areas extended into central Ontario and Quebec.

Passenger Pigeons were once so numerous that it would have seemed absurd to suggest that they were in danger of extinction. However, staggering levels of overhunting by professional hunters of these communal birds included massacres on the scale of daily killing of tens of thousands for months at a time at one of the last known large nesting sites.

The Passenger Pigeon went extinct in 1914, when Martha, the last known member of her species, died at the Cincinnati Zoo. Though attempts to breed them in captivity were unsuccessful, the legacy of the Passenger Pigeon is that their disappearance generated interest in strong conservation laws, and the strict enforcement of such laws has saved many other species from extinction.

The title of this piece was criticized for sounding awkward by a couple of people who did not catch the reference to the motto of the United States. *E Pluribus Unum*, translated from the Latin, means "Out of many, one."

Anatomy of the Lizards
January 8, 2013

Most biologists have done their share of dissections to learn about animal anatomy. Students typically dissect frogs, rats, sharks, and cats in comparative anatomy classes.

These exercises have tremendous educational value, but I just couldn't face them and fulfilled the requirement with a course called the Biology of Higher Vascular Plants. I developed a lifelong love of cosmos flowers, which is what grew from my mystery seed project. But I digress.

Perhaps it's because I never took an advanced vertebrate anatomy class that I find their basic structural anatomy so mysteriously fascinating. The group of species (or "taxon" to use the scientific term for a group of species all lumped together under a single name) whose anatomy most intrigues me is the lizards.

Their bodies definitely fall into the freak show category, which is actually a compliment, in case you thought otherwise. Consider the best-known anatomical peculiarity of some lizards: They can lose their tails and grow them back. The tail breaks off at a weak point in the vertebrae, which allows lizards to escape predators that have hold of their tail.

After separating, the tail provides a meal to the predator, and many predators abandon the chase for the rest of the animal. The detached tail will often move, which presumably keeps the predator's attention on it as the lizard flees for safety. The wiggling tail creates the illusion of a continued struggle, and that is more likely to occupy a predator than a body part that remains motionless. Lizards' new tails often grow back thinner and are sometimes different in color than the original, but that hardly detracts from the wonder that the tail can regenerate at all.

Besides being capable of autotomy (the self-severing or self-amputation of an appendage) and of growing a new tail, some lizards can lick their own eyeballs with their long tongues. It's a very handy way to cleanse the eye, which we humans tend to do with our rather less flashy blinking behavior.

Lizards don't just use their tongues for maintenance. They are really long (how else could they reach their own eyes?) and can be used to smell when stuck out. Lizards can even pull air toward the roof of the mouth where the Jacobson's organ can detect the chemicals in it and send messages to the brain about them.

Lizard legs extend out to the side of the body much like the position of human arms during the down part of a push-up rather than directly under the body in an erect stance. This may seem insignificant, but in related groups, such as dinosaurs, the erect stance allowed animals to breathe more easily while moving, which was critical for endurance and increased levels of activity.

The erect stance also likely promoted the evolution of larger body size because of the decreased amounts of certain types of stress on the limbs.

Lizard anatomy is so interesting that it's possible to make the case that these reptiles are as cool as cosmos flowers!

Satisfying the anatomy requirement of my UCLA undergraduate degree in biology with a botany course allowed me to dissect plants instead of animals and to learn a lot about those plants and many others.

Age-Related Division of Labor

January 22, 2013

Over the holidays, many families set up and then cleaned up elaborate dinners in the traditional way: Adults cooked, teenagers cut and chopped, young kids set the table. Young kids put small items back in the fridge, teenagers did dishes, adults organized leftovers. Everyone had a job, and the work was done efficiently.

If you are feeling smug about this well-oiled human machine and thinking how superior we are to other animals, cut it out right now. Many species partition jobs among different individuals. In fact, the strong division of labor within colonies of social insects is considered one of the reasons for the unparalleled success of these species.

An advantage to division of labor is higher efficiency. Individuals who specialize can become better at performing a task because they improve their performance through practice and because they continually receive information related to that task. Specialization also lessens the costs of changing tasks, which includes time and energy spent traveling between locations in order to perform a different task.

If you're hoping for human uniqueness, you'll have to look elsewhere because we're not even the only species with age-based division of labor. Honey bees, for example, who have arguably the most organized work force in the world, exhibit age-related division of labor, or as scientists call it, temporal polyethism.

Temporal polyethism is a mechanism that results in the allocation of tasks among workers based on their age. Though it also occurs in wasps and ants, it is best understood in the honey bee.

The youngest bees clean cells, a task they can perform despite the developmental immaturity that may limit their ability to fly and sting. Once they are a few days old, they begin taking care of the brood (the young, developing bees).

At the age of around 1 to 2 weeks, they are considered middle-aged bees, and they move on to receiving food from foragers, storing that food in the nest, maintaining and repairing the nest, and guarding the nest entrance. The final stage, which typically begins when bees are about 3 weeks old, is to forage. Once bees begin to forage, they no longer engage in tasks within the nest, but continue to forage until they die, which typically happens around 6 weeks of age.

Division of labor based on age is advantageous for colony survival and productivity. As bees age, the location of their work moves farther from the center of the nest, which minimizes wasteful movement and helps keep young workers safest. Foraging is a dangerous activity that exposes bees to predators and risks debilitating wear and tear on their wings. Bees only engage in this activity after contributing to the colony through their previous work, which maximizes the amount of work each bee performs during her lifetime.

Though honey bee temporal polyethism is an impressive mechanism of division of labor, it's just as well that they aren't for hire. In spite of their efficiency and organization, it's hard to imagine that having honey bees cater and serve at your next family gathering would work out well.

My advisor in graduate school, as well as several of my fellow graduate students, studied temporal polyethism in social wasps.

Crime-Fighting Insects
February 5, 2013

A Chinese peasant was killed by repeated stabbing, suggesting that the murder weapon was a rice-harvesting sickle. The lawyer and death investigator, Sung Tzú, summoned all the suspects, requesting that they bring their sickles. Sung Tzú ordered them to put their sickles on the ground and step away from them. Over the next few minutes, shiny metallic flies began to collect on only one of the sickles. With the evidence of the flies against him, the owner of that sickle confessed to the crime.

The flies provided such strong evidence against the murderer because they were a specific type of blow fly that is attracted to blood. Though the murder weapon was cleaned off enough to fool human eyes, the flies still detected the scent of blood on it.

This is the first known case of solving a crime with forensic entomology, which is the use of insects (and their arthropod relatives) to aid legal investigations. It took place in the 1200s, making it obvious that forensic entomology far pre-dates *CSI*.

One key way that insects help investigators solve crimes is by determining the time of death. After 24 to 48 hours, most estimates of time of death are not very good, but insect data remain reliable. Many insects are attracted to the dead, especially blow flies and flesh flies. Blow flies often colonize the body within minutes, and sometimes even seconds, of death. If a scientist knows the species of insect and the stage of development, it's possible to obtain a very accurate estimate of time of death.

It's not uncommon for criminals to set fire to a house after a murder to make it appear that the person died in the fire. When insects on the body are more advanced in development than expected, based on the time of the fire, it is strong evidence that the person died before the fire and that someone was trying to cover up the murder with fire. The absence of any insects may mean that a body was frozen, sealed in a tight container, or deeply buried.

In another situation in which insect evidence was valuable, the prosecution claimed that the suspect drove a rental car from Ohio to California and killed five people there. The defense argued that the car had never left the Ohio area. Several insect species picked from the car parts are found only in the West and one was abundant in California. The entomological evidence was consistent with two major paths to California from the suspect's area and the fact that there were 4,500 unaccounted-for miles on the rental car.

Insect evidence sometimes saves innocent people from being accused. In one case, the investigator determined that the assemblage of insects present indicated that a body found in the wall of an apartment had been there since before the new people moved in. This shifted suspicion away from them and onto the previous tenants.

Forensic entomology frequently provides evidence that allows investigators to solve crimes. It's strangely satisfying to know that many criminals can't even outsmart insects.

Knowing that insects help prosecutors secure convictions against criminals is gratifying in terms of justice, and also as a reminder of both the importance of science and those with scientific expertise.

Birds Learn from One Another
February 19, 2013

In 1921, songbirds began to irritate people by eating the cream that collected at the top of milk bottles that were delivered to their porches. Most annoying is that this "bird problem" was one that people thought was solved. In the past, milk bottles had no sealed tops, so the cream was easily accessible, and birds of many species drank from them.

Later, aluminum seals were placed over each bottle, and for a few years, people had the upper hand. Then, in Swaythling, England, birds were at it again, pulling off the foil caps or piercing them. Either method allowed them to consume the cream (a rich and valuable food resource) and aggravate their human neighbors. Over the next few decades, people observed birds doing the same thing throughout the U.K., in Ireland and in many places within continental Europe. How did this behavior originate and how did it spread so far?

The answers lie in an understanding of both cultural transmission and the foraging patterns of the species that did most of the cream stealing—the Blue Tit. (If you think that's the funniest name for a bird species, then you've never heard of the Great Tit or the Elegant Tit.)

Cultural transmission is the transfer of information to others through social channels. Socially transmitted and learned information, as opposed to what's passed on genetically, helps form the culture of a species, and individuals rely on experience and participation to learn. Blue Tits are social and often forage in groups, so there were ample opportunities for naïve birds to observe experienced birds gaining access to the cream. The result is that the behavior of opening milk bottles spread.

The foraging behavior of Blue Tits predisposed them to discovering how to break into milk bottles. They are inquisitive and drawn to bird feeders and other foods provided by humans. One of the ways that they search for the insects they eat is by poking at bark or peeling it away, and this behavior is similar to what they do to milk bottles.

The spread of the milk bottle opening behavior followed the pattern usually observed in culturally transmitted behaviors, including its spread over a large geographic area. There was an initial slow period of growth followed by a rapid spread. Usually the next phase is a slowing down of the spread as the number of birds not already performing the behavior diminishes. However, the study of this behavior stopped prematurely in the mid 1940s, and there were still many regions with birds who had not yet been exposed to this behavior.

There was no consistent relationship between the distance from Swaythling and when the behavior appeared in an area, suggesting multiple instances of individual birds figuring out how to get past the foil covering to reach the cream. Because of their curiousness and natural feeding behaviors, it is not surprising that there were many innovators among Blue Tits.

Blue Tits passed on their milk bottle opening know-how without tweeting, but it's still a story about social networks.

The cream-stealing Blue Tits are close relatives of our chickadees.

Not Extinct After All
March 5, 2013

If no scientist had ever seen a living specimen of an animal, and if the most recent samples in the fossil record date back 65 million years, then considering the animal extinct seems like a safe bet. However, in one case, even that extraordinary gap turned out to be misleading. A lucky series of events revealed to the scientific community that there are still living coelacanths (pronounced SEE luh kanths).

Coelacanths are lobe-finned fishes, of which more than a hundred species are known from fossils. They were presumed extinct until 1938, when a fisherman named Hendrik Goosen caught one off the coast of South Africa. When Goosen docked at a small port in late December, he contacted Marjorie Courtenay-Latimer, who was the curator of a museum in East London, South Africa. She sometimes purchased unusual fish for the museum.

Latimer was extremely busy, but went to the dock anyway to give holiday greetings to the ship's crew. When she saw the coelacanth, she knew it was something unusual and she wanted it for the museum even without knowing for certain what it was. She described it as "the most beautiful fish I had ever seen, 5 feet long, and a pale mauve blue with iridescent silver markings." Her first challenge was to convince a cab driver to allow the large, stinky fish into his vehicle so she could transport it to the museum.

With that hurdle crossed, she preserved the specimen as best she could without refrigeration and contacted a professor at Rhodes University in Grahamstown named James Leonard Brierley Smith, who immediately recognized it from her description and drawing. He came to East London to see the specimen himself, knowing that he would be thought a fool if he erroneously announced the existence of a coelacanth. One look at the fish convinced him that he had been correct in his identification. As the internal organs were lost in the mounting process, Smith was beyond eager for the opportunity to study a complete specimen.

It was not until 1952 that Smith was able to see another coelacanth, when one was caught in the Comoros Islands. If it had not been for the flyers Smith had distributed all along the African coast as far north as Kenya, this fish would have been lost. The fisherman was about to clean it when someone informed him that there was a reward of 100 pounds available for it.

Fishermen in the Comoros Islands were familiar with this rare fish, which they called *gombessa*, but since it was not edible, they were not overly interested in it. So, while scientists didn't know that coelacanths still lived, these fish had been seen occasionally by people prior to the 1938 discovery that shook the scientific world.

To say that the scientific world was shaken is no hyperbole. Seeing a living coelacanth was as surprising as it would have been to see a living velociraptor or a triceratops or any of the many dinosaurs that stopped appearing in the fossil record at the same time as the coelacanths.

Only two species of coelacanths are known to be alive today, though 120 are known from the fossil record. In addition to the species found off the coast of South Africa in 1938, there is an Indonesian species that was first seen in a fish market in Indonesia in 1997.

Bioluminescence
March 19, 2013

To describe the first time he observed bioluminescence during a dive, my scuba instructor revealed: "If I had been a biologist, I would have reached a transcendent state that would have made nirvana seem ho hum, but I'm a computer scientist, so I was merely moved." He was preparing our advanced class for a night dive during which we expected to see this phenomenon, which is truly one of the coolest experiences in nature.

Bioluminescence is the production and emission of light by a living organism. The word has a dual origin, being a hybrid of the Greek word for "living" (bios) and the Latin word for "light" (lumen). The light comes from a chemical reaction that produces "cold light," meaning that nearly all of the energy created by the chemical reaction goes into producing the light rather than being released as heat.

Throughout the living world, many different chemicals are used to produce light, which is why scientists believe the ability to give off light evolved multiple times in different lineages. This idea is supported by the fact that the phenomenon occurs throughout the living world. In addition to some animals, many bacteria, fungi, and protists are capable of bioluminescence.

In the animal kingdom, the species capable of bioluminescence include fireflies, anglerfish, centipedes, click beetles, squids, cookie cutter sharks, coral, jellyfish, clams, krill, snails, eels, and starfish.

Bioluminescence has many purposes: mate attraction, territorial displays, communication, defense, mimicry, and predatory functions.

Despite the widespread existence of this phenomenon in animals, the most spectacular displays of bioluminescence come from marine plankton called dinoflagellates, which are protists.

One of the most extraordinary places to observe dinoflagellates light up is in the brightest bioluminescing bay in the world: Bioluminescent Bay on Vieques, a small island off the east coast of Puerto Rico. Nearly three-quarters of a million bioluminescent organisms are in each gallon of water, and they glow with a blue-green light when agitated. My family was there on a moonless night a few years ago, and as we kayaked into the bay and swam in it, each motion of paddle, boat, or body created an explosion of lights in the vicinity. Every time the water was agitated, it looked like fireworks had been set off.

Dinoflagellates use bioluminescence as a defense mechanism. The flash of light may startle predators, and the short delay in their predatory actions may prevent them from successfully eating the dinoflagellate. It has also been hypothesized that the light produced by these microorganisms may function like a burglar alarm, illuminating the predator to draw attention to it, which might make it vulnerable to its own predators.

When you are in the water on a dark night in an area with many bioluminescent microorganisms, the bright lights are even more astounding than the best starry night viewed from our International Dark Sky City of Flagstaff. At least that's what I think as a biologist. An astronomer would probably disagree, so we may need a computer scientist to break the tie.

Most bioluminescent organisms live in the ocean, and the best places to see them are shallow bays with narrow entrances to the ocean so they are protected from waves. They appear most spectacular on dark nights with a new moon.

Like Flies to Alcohol

April 16, 2013

If you give a male the choice to consume alcohol or to abstain, his decision will be influenced by his recent success or lack thereof with females. If females have been receptive to his advances, he is more likely to pass on the alcohol, but if he has been rejected, he is more likely to turn to alcohol.

It's easy to imagine a male drowning his sorrows with a stiff drink or two after being rejected by the female of his choice. Such a response is hardly startling, so the research confirming this behavioral phenomenon is no shock, either. The revelation that the males and females that were the subjects of this scientific investigation were all fruit flies, however, may come as a surprise.

A recent study found that male fruit flies who had been rejected by females were more likely to choose food with a high alcohol content over food without alcohol when compared to males who had successfully mated. In the experiments, males were either put into containers with virgin females with whom they then mated, or in containers with females who were already mated and therefore no longer receptive to males.

A series of experiments allowed researchers to determine that it was sexual deprivation itself rather than the social experience of rejection that led to the increased consumption of alcohol. If males who had been deprived of mating opportunities originally were later allowed to mate, their consumption of alcohol declined.

This is the first study to show the impact of a social interaction on future behavior in fruit flies, and investigators followed up these initial studies by looking into the mechanisms behind the choices males made with regards to alcohol consumption. They looked at the levels of NPF (neuropeptide F) in the flies. A similar chemical, called neuropeptide Y (NPY), mediates alcohol consumption in mammals including humans. Both NPF and NPY are chemicals in the brain that function in reward systems. They are among the chemicals that make animals of many different species feel good, signaling to their bodies that they've probably just done something that is evolutionarily advantageous, such as eating or mating.

Male flies who successfully mated had much higher levels of NPF than males who had been rejected, so they did not consume alcohol to raise its levels. Researchers have postulated that experiences that change the levels of NPF in flies influence their behavior because there is a tendency for the flies to behave in ways that restore NPF to normal levels. Sexual deprivation results in lower NPF levels and leads flies to take actions that increase NPF levels, such as consuming alcohol. Successful mating elevates the level of NPF, resulting in flies that aren't looking for ways to raise it.

Practical applications of this study lie in its potential to improve treatment of addiction in humans. The work increases our understanding of the physiological need for reward centers in the brain to be activated. It also leads to a whole new meaning for the term "barflies."

Of course the similarities in behavior between different species is fascinating, but what really piques my interest in studies such as this is the way the same physiological processes influence behavior in animals that are not closely related.

Animal Architects
May 21, 2013

The Taj Majal, the Sydney Opera House, the Eiffel Tower, the Empire State Building, and the Egyptian pyramids are wonders of human architecture. We rightfully take pride in these structures, which were designed and built by some of the brightest and most creative members of our species. In other species of builders, even the most ordinary individuals are true architects.

Architecturally speaking, the large scale and environmental impact of beaver dams stand alone. To begin, beavers divert the stream to decrease water flow. Then they place logs and big branches into the mud to make the dam's framework. Sticks, leaves, rocks, mud, and grass supplement the basic structure, adding to its size and strength. Beavers build these structures to an average height of 6 feet and a width of 5 feet. Dams are angled to resist the current and have spillways allowing excess water to pass without causing damage. By preventing the free flow of water in waterways, beavers make ponds, meadows, and wetlands, creating habitat for fish, frogs, and birds.

For finesse and complexity, the trapdoor spider's architectural accomplishments are hard to match. These spiders build burrows with a trapdoor entrance, from which they get their name. The trapdoor is made of plant matter, silk, and soil, resulting in a structure that is similar to cork. The hinge is made of silk. It's difficult to see the trapdoor when closed because its materials camouflage it so effectively. These spiders can secure the door from the inside by holding it with their pedipalps (small appendages near their mouths) to protect themselves against predators. Trapdoor spiders are nocturnal, sit-and-wait predators that hold onto the underside of the trapdoor from within the burrow. Pieces of silk arranged outside the burrow entrance act as trip lines. Vibrations alert the spider to the presence of prey. When it is close enough, the spider pops out of the burrow to capture the prey and make a meal out of it.

Perhaps it's because most of my research has been on wasps that I think the architectural achievements of paper wasps are among the most impressive in the animal world. For one thing, they have been making paper for millions of years longer than humans, who only started doing it roughly 2,000 years ago. Paper wasps combine secretions from their own body with plant fibers to create paper that they use to make complex nests. The nests have cells to rear their young, and many of them also have a covering, called the envelope, which protects against parasites. Depending on the plant fibers used, nests can vary in color. Some nests are big and resemble rounded cardboard boxes the size of exercise balls, and others look just like little cupcakes. Among the tiniest ones are those built on the underside of a leaf and made to look like a dying leaf, complete with faux veins that match the actual leaf supporting the nest.

Now that's the kind of organic architecture that even Frank Lloyd Wright could admire.

It's so often the case that animals preceded us in success with so many endeavors we like to think of as very human.

Social Cooking

June 4, 2013

Being social allows organisms to accomplish things that would be impossible alone: Wolves and orcas hunt in packs so they can take down larger prey than a single animal could. Ants find every food source (including all picnics) within a large radius of their nest. Many pairs of eyes provide constant vigilance against predators in animals as varied as meerkats, squirrels, sparrows, and gazelles.

Honey bees have maximized many of the benefits of sociality, including better temperature control and moisture retention, as well as an advanced task specialization (also called division of labor) that enhances efficiency. On the downside, groups of organisms do attract predators.

Between the honey stores and the large number of succulent larvae inside, a true bonanza awaits those who can raid a honey bee hive. Many honey bee predators wouldn't bother if there were only a single larva and a drop of honey to entice them.

One of their most dangerous natural enemies is the Asian giant hornet, which is the largest hornet in the world. Individual honey bees are defenseless against this hornet. They can't sting the hornets because honey bee stingers are too small to pierce through a hornet's exoskeleton. An Asian giant hornet, with a wingspan of 3 inches and a 2-inch body, is roughly five times larger than a honey bee. If a hornet discovers a honey bee hive and returns to her own colony, she can lead more hornets to the honey bee hive.

Once a giant hornet has recruited other members of her colony to raid it with her, there is little the honey bees can do to defend their hive. The group of hornets will completely overpower the honey bees, eat all the young living inside, and kill a lot of the adult bees as well by tearing them into pieces at a rate of several dozen per minute.

However, if the bees can prevent the first hornet from returning to her nest to recruit help, they can keep their own hive safe. Japanese honey bees are capable of killing hornets that arrive in small numbers. They respond to this serious predatory threat by working together to accomplish what would be impossible for a single individual.

Japanese honey bees form a "hot defensive bee ball" in which hundreds of bees engulf an invading hornet and vibrate their flight muscles to generate heat. Within 5 minutes, the bees can heat a hornet up to a lethal temperature of 46 degrees C (114.8 degrees F). Hornets die at temperatures of 45.7 degrees C, but honey bees can survive temperatures up to 50 degrees C (122 degrees F).

Only Japanese honey bees can defend themselves this way. The European honey bee, which is a different species, cannot. Attempts to raise the highly productive European honey bees in Asia were abandoned because too many colonies were raided by the Asian giant hornet. European honey bees have not evolved a defense against them that is anywhere near as effective as the Japanese honey bee's social cooking behavior.

These giant hornets were called "murder hornets" by many in the media in 2020 and beyond. Those same sources failed to mention that their victims are honey bees, *not* humans.

That's How Dung Beetles Roll
July 2, 2013

With a world full of beautiful and charismatic insects to revere, it's not obvious why the dung beetle was the one considered sacred by the ancient Egyptians. Dung beetles don't seem sacrosanct by modern standards. They eat dung, preferring herbivore excrement to that of omnivores, and rear their young on it. They make dung balls that they roll to a soft spot in the ground and bury. Next, they mate underground and the female lays eggs in the dung.

The Egyptians hieroglyphic for the dung beetle means "to transform" or "to come into being." This insect represented renewal, rebirth, and resurrection, which is why it appeared in funerary art. The dung beetle's sacred status stemmed partially from a misconception about their basic biology.

The ancient Egyptians believed that all dung beetles were male and that they reproduced by depositing semen into dung balls. This generation of beetles from the self was likened to the actions of Khepri, a subordinate of the sun god Ra. Khepri was the god of the rising sun, who came into being out of nothing and who renewed the sun each morning before rolling it across the sky. The similarity between Khepri moving the sun through the sky and dung beetles rolling a ball of dung also explains the association between the two.

If you're tempted to snicker at these errors, you may also be amused by more modern confusion about dung beetles. In the late 1800s, the famous scientist Jean-Henri Fabre demonstrated that the "helpers" of other beetles rolling balls of dung were actually trying to steal the dung ball, *not* offering assistance. It's advantageous for dung beetles to move their balls of dung away from the source pile quickly to get away from thieves. Hardworking beetles' time and energy will be wasted if they can't roll fast enough to avoid being robbed. In order to make a quick getaway, dung beetles typically move in a straight line when rolling dung.

For many years, scientists have observed dung beetles rotating in circles while on top of the dung ball, doing what appeared to be a dance. Earlier this year, it was reported that these impressive displays of balance and coordination allow the beetles to navigate in the straight lines that are critical for preventing theft. The movements of the "dance" allow them to reorient if they fall off the dung ball or are otherwise at risk of going off course.

Their navigational system is based on celestial cues such as polarized light from the sun. They can also orient themselves using the weaker light from the moon, but researchers were initially puzzled by beetles' ability to roll dung in perfectly straight lines even on moonless nights. Further study revealed that the beetles use the bright strip of light from the Milky Way to navigate, making them the only animals other than humans who are known to do so.

I guess the ancient Egyptians were wise to connect the lowly dung beetle to the wonders of the heavens after all.

I resisted the urge to include this joke in the original column: a dung beetle walks into a bar and asks, "Is this stool taken?"

Language Is Crawling with Insects
July 16, 2013

A bug can refer to a computer glitch or a device for spying. Being cute as a bug is good, but to bug someone is bad. It's also bad to use "bug" as a synonym for insect. Bugs are only those insects in the suborder Heteroptera, including seed bugs, giant water bugs, and stink bugs. Insects that are not bugs include bees, beetles, butterflies, fleas, grasshoppers, ants, crickets, flies, and cockroaches.

Language related to insects may be imprecise, but it is abundant. You can be a fly on the wall or a fly in the ointment, be as busy as a bee, have ants in your pants, a bee in your bonnet, or butterflies in your stomach. You can stir up a hornet's nest, be bitten by the love bug, or attend a flea market. You can say, "Don't let the bed bugs bite" or that someone "wouldn't hurt a fly."

Some insect terms are surprisingly technical. To make a beeline means to take the direct route from one point to another. Bees with a full load of nectar return to their hive in a straight line from where they finish foraging. Many people think a beeline refers to the path a bee takes from the hive to a flower, but bees typically take highly circuitous routes to flowers, visiting many on a single trip and only then flying straight home—on a beeline. That's a huge navigational feat for insects foraging miles away from the hive.

The butterfly effect is a phenomenon in chaos theory, referring to the extreme sensitivity of some systems to initial conditions. A small difference at the beginning in certain systems can result in a huge difference in the conditions later on. It most commonly refers to weather, where sensitivity to the smallest of changes in conditions makes weather predictions more than a few days into the future so unreliable.

This effect refers to the idea that a flap of a butterfly's wings creates tiny atmospheric changes that could change the path of a storm, or delay it, or cause one to form elsewhere. The wing flaps don't directly cause a hurricane or a tornado by providing energy to the system. The butterfly just alters the initial conditions, which causes a chain of events that leads to large-scale differences later on.

The flea flicker is a trick play in football, a pass play that is designed to look like a run to the defense. The quarterback hands off (or laterals) to another player who pitches it back to the quarterback. The quarterback then passes to one of the receivers who is wide open because the other team stopped defending against the pass in order to cover the run. The flea flicker is a play that has been around since you were knee-high to a grasshopper and it's the bee's knees. It leaves the fans bug-eyed, draws the offensive players to one another for high fives like moths to a flame and makes the defense as mad as a hornet.

Insects have long been important in human society because they matter for both our survival and our comfort, so it is no surprise that they figure prominently in our language.

Power of the Mob
August 20, 2013

"Wow, I've read about this *%!*&!" shouted a fellow classmate during a field trip with our animal behavior class. He was pretty excited to see blackbirds mobbing a hawk, and I shared the feeling. Mobbing behavior is an example of the triumph of the little guys over a larger, more powerful individual. As a student of social behavior fascinated with the benefits of cooperation within groups, I never tire of observing it in real life.

Mobbing is an anti-predatory behavior in which a group of animals harasses a predator to make it leave the area. When smaller birds mob a predator such as a hawk, they dive bomb the larger bird, squawk at it, fly around it relentlessly, and even defecate or vomit on it. Interestingly, they are able to eliminate on their enemies with considerable accuracy in flight—an enviable skill. On occasion, predators have been unable to fly because of the extra weight of the feces or vomit.

Birds that are mobbing often give mobbing calls. These calls serve to attract other individuals in the vicinity to join the efforts to repel the predator, and are different from alarm calls in both sound and function. The purpose of alarm calls is to alert other members of the species to the presence of predators so that those so alerted can escape the predator, not head toward it to actively repel it from the area, as mobbing calls cause individuals to do.

Any behavior in which a group of individuals is around a potential predator is typically considered mobbing, and such cooperative attacks and harassment can effectively deter predators. One of the ways that mobbing provides protection against predators is by eliminating the surprise attack, which is one of their greatest weapons. Once a predator has been noticed and its presence communicated to other individuals with calls and the conspicuous mobbing behavior itself, its chances of success drop considerably.

Mobbing behavior is most prevalent during the breeding season. Species whose young are vulnerable to predation are most likely to engage in this behavior. Many gulls, terns, and kittiwakes nest in colonies of up to thousands of pairs of birds on islands or on sea cliffs and they mob any predators that approach. In forests during breeding seasons, playing recorded calls of chickadees that are mobbing owls brings responses from roughly 50 species of birds, allowing scientists to determine which species are present far more rapidly than other survey methods.

Mobbing is most often described in birds, but occurs in other species, too. There are squirrels that mob snakes, and African buffalo work together to mob lions. Some fish have even been observed mobbing turtles to protect their young. Meerkats mob two of their most dangerous predators—snakes and foxes—with behaviors like poking, growling, and biting.

Besides deterring predators, mobbing teaches the next generation which species are dangerous. Populations that are reintroduced into the wild sometimes fail because they lack knowledge about the identity of important predators. Education matters in all species!

Where I live in Flagstaff, Arizona, the most common type of mobbing I see is American Crows harassing Red-tailed Hawks.

In the Beginning, There Were Termites

September 3, 2013

I was running as fast as possible through a Costa Rican rain forest toward the beach because white-faced capuchin monkeys were chasing me. Worse, they were doing what primates do to unwelcome intruders, which is throwing feces at them.

The ocean offered the best refuge, so when I reached the water, I plunged in. That moment marks the start of my life in science. My first thought was not, "Wow, that was an experience that will stay with me forever," or even "Yikes, scary!" It was, "Darn, that's an entire trial in my experiment ruined."

I had been working on my independent research project for the field studies course I took as a college sophomore. My study involved termites' ability to distinguish nestmates from individuals of the same species that are from a different colony. Amazingly, in colonies of many thousands of individuals, these soldiers and blind workers can tell the difference. They pass by nestmates without reacting, but soldiers attack and workers flee from individuals belonging to any other colony.

The mechanism of this recognition is based on the termites' sense of smell. They can identify nestmates and distinguish them from non-nestmates because every colony has a distinctive colony odor. The particular pattern of hydrocarbons on the outer surface (exoskeleton) of all colony members is responsible for the unique smell.

Before the monkey incident, I had been introducing either nestmates or non-nestmates to colonies of termites and observing the response. Did they attack the termite being presented or accept it into their colony? I had put a termite into the nest and nearly completed the observation period when I was distracted by flying poop and the distressingly loud vocalizations of a troop of extremely pissed-off monkeys. (That's not a technical description that you'd find in the scientific literature, by the way.)

Now, 25 years and six trips to Costa Rica later, I'm back where it all began. I'm in Costa Rica discussing with my sons the same type of termites that launched my interest in social insects that led to doing a PhD on tropical social wasp behavior.

Termites in the genus *Nasutitermes* are easily identifiable by the unusual soldiers. Most termite soldiers have large mandibles for defense, but these termites have snouts from which they can spray noxious and sticky chemicals. These turpentine-like chemicals trap insect predators and deter anteaters by being toxic and foul-tasting.

Like other social insects, the majority of the colony members are sterile workers. Kings and queens reproduce while the rest attend to feeding the young, repairing the nest, foraging, defending against predators, and everything else besides laying eggs and fertilizing them.

Alates (also known as reproductives) are the only termites that have wings. When colonies produce alates in April or May, there are thick clouds of thousands of flying termites leaving the nest in search of mates. These swarming termites are so dense that I later dreamed that their presence saved the day, blocking the monkeys' view of me and preventing the disruption to my experiments.

Our family spent the fall semester of 2013 in Costa Rica when our kids were in third and fourth grade, and it was during this time that I renewed my love of termites.

Frogs of My Childhood

September 17, 2013

My parents differ from many other animal lovers in that they don't want to have them in their house. They love to see all types of wild animals around the world in the course of their travels. They're in favor of environmental policies that protect endangered and threatened species. Most important from my perspective, they support both my love of and my work with them.

But their preference for a certain separation between their home and those of other species meant that I didn't have many pets growing up. Until Mittens the hamster when I was 15, my menagerie was limited to goldfish and frogs.

At age 9, I had four pet frogs, only two of whose names I can recall. Leprechaun Jumper and Gnome Mobile were about an inch long, while the other two (now nameless) were half that size. They were captured by a creek in Southern California back when that wasn't a crazy thing to do. (I'm officially a geezer since I'm on the verge of referring to the good old days.)

My best guess based on comparing my memory to pictures from field guides of Southern California amphibians is that they were California tree frogs.

California tree frogs are well camouflaged, being mottled gray and brown with even darker blotches in some areas, and lighter on their undersides. They are bumpy along their backs, and have notable webbing between their toes as well as enlarged toe pads. They're hard to spot in their habitat of rocky streams, but catching them isn't a great challenge. Since they don't struggle much, they're easy to hold without causing injury.

Like most pets, these frogs had their advantages and disadvantages as members of the household. On the upside, it was easy to take care of them. We kept them in a 10-gallon container that had a pond, a raised area where the frogs could rest, and vegetation. We fed them flies that we collected around the house and yard and put in the container live for the frogs to catch and eat.

On the downside, we lost them around the house more than once. Frogs (here's a news flash for you) can jump exceptionally well, which allowed them to leap out of our hands. They could go far quite quickly with just a few jumps, and because they are small, the possible places where they could hide might as well have been infinite. Luckily, we always retrieved them eventually, but there were times we felt great doubt that we would.

Because they were clearly eager to escape the container (and us), we let them go after a fairly brief visit with us. Having kept them for the better part of a month, we released them at the same spot where they were collected, hopefully not much worse for the wear.

It turns out that my parents and the frogs were kindred spirits, being in agreement that it didn't make a lot of sense to have these animals in our house.

I now know that collecting wild animals and keeping them as pets, even briefly, is a no-no, but I didn't know that when I was little. I have fond memories of my pet frogs, and hope that I caused them no serious harm.

Biting, Spitting, Leggy Marvels
October 1, 2013

"It bit me!" my son yelled from the yard, and I flew out there with the speed that only Mommy superpowers can grant. My kids had been observing a neighborhood cat that hangs around our birdfeeder, and I feared the worst. As a canine aggression specialist, I've often had reason to feel grateful that it was a dog and not a cat who had bitten my clients. Both types of bites can cause damage, but the infection rate of cat bites is roughly 10 times higher.

When I learned what actually happened, my previous moment's terror seemed ridiculous. Yes, an animal in our back yard had bitten him, but it was not a cat, nor any other creature with teeth. It was a grasshopper, and though it hurt a little, my son was more interested in the event than injured by it.

When they are not biting my children and giving me an adrenaline rush, I'm quite fond of grasshoppers. They're such amazing packages of interesting biological oddities. Rather than have their ears on their faces as we might expect, they are on the legs—those same limbs that give them such spectacular jumping abilities. Grasshopper legs work much like a catapult. The powerful hind legs contract, and then relax, at which point a special piece of cuticle at the bend in the leg releases energy like a spring and thrusts the grasshopper into the air. If people could jump as far as these insects can relative to body size, we would routinely be soaring the length of soccer fields.

The wonder of their legs does not stop there. Grasshoppers are also capable of stridulating (pronounced STRID-yoo-lay-ting), which is producing sound by rubbing body parts together. The hind leg has a scraper with an edge to it and the grasshopper rubs that scraper against the forewing, which has fine ridges. In addition to these musical structures on the legs, grasshoppers have spines there, too, which help them defend against potential predators.

Besides the spiny legs and the ability to bite, grasshoppers have other ways to protect themselves from being eaten by the many birds, rodents, lizards, spiders, and snakes that would greatly enjoy having them for dinner. For one, they tend to be in large groups, which means that even if predators are in the area, each grasshopper has a small chance of being hunted because so many other individuals dilute the risk. Individually, one of their last-ditch efforts to protect themselves from being eaten is to spit. If they are caught, they exude a slightly acidic substance from the mouth that is made of partially digested food. This form of defense relies on the "ick" factor, meaning that its goal is to repulse an attacker enough to drop the grasshopper rather than to eat it.

If only my son had yelled "It spit on me!"—instead of "It bit me!"—I would have known that a grasshopper was responsible, and I would only have been grossed out rather than freaked out.

I was recently walking around a nearly dry lake near my house and the place was hopping with thousands of grasshoppers. Though many other animals were taking advantage of this abundant food source, their high numbers meant that the majority would survive.

The American Crocodile
October 15, 2013

Few animals are as scary to most people as crocodiles. Not counting the sharks, no animals are called "man eaters" with greater frequency. Crocodiles are indeed formidable predators. This partially explains the evolutionary success of this ancient lineage, which evolved at about the same time as dinosaurs but has changed little over the last 100 million years.

The weapons of the American crocodile include large sharp teeth that are suitable for holding onto their prey rather than for tearing its flesh, and jaws with enough power to make those teeth bite down with a force of 3,000 pounds per square inch. These animals can also inflict damage with sharp claws. They specialize in attacking by ambush, meaning that they sit and wait for a meal to come close and then attack. The majority of their food comes from small prey such as insects, fish, birds, turtles, frogs, snails, and crabs. They only occasionally eat cows and horses, and rarely attack humans. What a relief!

American crocodiles can swim 20 miles per hour over short distances, with a streamlined body and the ability to tuck their feet in to lower water resistance. They use their webbed feet to maneuver in the water, and for such big and bulky animals, they are remarkably agile.

Besides their predatory talents, they have the ability to live in salt water. The American crocodile is more than salt tolerant, actually preferring to live in the salty or brackish water of coves, canals, rivers, and coastal mangroves, where they set up territories and defend them from other crocodiles.

Even though they are strong and resilient, they have their weaknesses. They require warm water, and are so sensitive to cold that they can quickly become helpless and drown in water that their close relatives, the alligators, can tolerate. A 2010 cold snap in Florida killed over 100 wild ones, and it's their inability to handle lower temperatures that keeps crocodiles from thriving north of the tip of Florida. Though their habitat extends from Florida and many islands of the Caribbean to as far south as Peru and Venezuela, they are especially abundant in Costa Rica.

On a recent boat tour on the Tárcoles River in Costa Rica, I saw many American crocodiles. Though we saw a few around 15 feet long, it was shocking to ponder that they are not the biggest crocodile species, an honor which belongs to the saltwater crocodile of Asia and Australia. Individuals of that species can be 23 feet long and exceed 2,000 pounds. Male American crocodiles weigh in at 850 pounds while females are less than 400 pounds. Though small relative to saltwater crocodiles, American crocodiles are larger than Orinoco crocodiles, Cuban crocodiles, and Morelet's crocodiles, making them the largest crocodile species in the Americas.

Crocodiles scare me and interest me in equal measure. Viewing them in the wild is a rush, but it's hard to tell whether the thrill is fear or fascination, because they truly are as terrifying as they are enthralling.

Though our family took a boat ride on the Tárcoles River, it's easy to see them from the pedestrian walkway on the famous crocodile bridge over this river. Everyone driving from Jacó (a beach with great waves) south along the west coast of Costa Rica crosses this bridge.

Crowds, Cues, and Wildlife
November 5, 2013

We looked eagerly along the beach in both directions hoping to see a crowd of vultures either standing on the ground between the forest and the water's edge or flying around. One of the best ways to find a nest of sea turtle hatchlings heading to the ocean is to spot the birds that are making a meal out of them. Using the presence of one animal species to help locate another species is very common, and we used it to our benefit in Costa Rica's Tortuguero National Park.

To be honest, we were alerted to the presence of hundreds of newly hatched baby turtles making a break for the ocean by a crowd of people on the beach rather than a crowd of vultures, but the principle is the same. And this principle helped us see not only turtles, but other interesting species as well.

One of the best finds was a Double-toothed Kite that our son Brian pointed out to us. He has found a number of the birds of prey that we observed, having a real knack for spotting them. In this case, his knowledge of the bird made all the difference. He had read in our field guide that this species is often found in the company of white-faced capuchin monkeys. While the rest of us were excitedly watching a troop of these primates forage along the edge of the river near our boat, he was searching the area for kites, and he was not disappointed.

It's far easier to spot the monkeys than the Double-toothed Kite. The monkeys forage noisily in social groups, moving the branches of trees wildly and generally making their location obvious. In contrast, the bird is more secretive. It's solitary and on the small side for a raptor. It spends a lot of time perched motionless near troops of monkeys waiting for their movements to flush the lizards, insects, and other small animals that make up much of its diet. It then swoops down on its prey. Trying to find one of these birds in the forest where there are endless branches, trees, and thick foliage with which it can blend in so easily is a daunting task. However, when the search is narrowed to the immediate area of a troop of foraging white-faced capuchins, the chances of seeing one rise considerably.

Some animal species serve as a warning of what's nearby rather than as a clue to a potentially breathtaking wildlife sighting. When you see a swarm of dragonflies dipping, weaving and swooping through an area, it's likely that they are there because of the mosquitoes. When conditions are favorable for mosquitoes, they can be present in such large numbers that their swarms have been called "bugnadoes," and dragonflies feast on them.

If you seek wildlife, then looking for groups of vultures, monkeys, dragonflies and even humans may inform your search in accord with the words of A.E. Coppard: "To be far from the madding crowd is to be mad indeed."

Being able to spot animals makes time spent outdoors that much more fun, and the knowledge of where and how to look makes finding interesting species more likely.

Flying Animals That Can't Fly

November 19, 2013

If you're trying to determine whether an academic discipline is a science, don't base your decision on seeing the word "science" in the name. Biology and physics are scientific disciplines, but political science and military science are not. Similarly, don't assume that an animal can fly because its name includes the word "flying."

The only flying animals living today are birds, insects, and bats, and their names don't mention it. Among the animals whose monikers erroneously suggest that they have this ability are flying squirrels, flying fish, flying snakes, flying frogs, flying squids, flying lizards, flying geckos, and flying lemurs.

Unless animals use muscular power to generate the aerodynamic forces that allow powered flight, they aren't truly flying, and are unable to maintain their altitude or speed. They can, however, glide through the air and go long distances before landing. It's because of their spectacular gliding abilities that they are called flying animals, even though the term isn't technically correct.

It's advantageous to fly because flight offers new ways to escape from predators, new places to breed, new food sources, and the possibility of long-range migration to maximize the exploitation of seasonally available resources. Being able to fly no doubt contributed considerably to the evolutionary success of insects, bats, and birds.

Even gliding provides tremendous advantages, which is probably why it has evolved at least a dozen times and why there are so many different types of animals that can do it. Gliding animals tend to jump from high places, such as trees, and descend continuously while also moving horizontally. (Gliding is the movement downward through the air at an angle of less than 45 degrees from the horizontal.) Drag and lift prevent gliders from going into a free fall, with the necessary lift provided by membranes. Different parts of the body form the membranes in the various "flying" animals.

Flying fish swim very fast, and then burst out of the water, using their enlarged fins to stay airborne for 50 meters or more. Flying frogs are able to glide because of their enlarged toe membranes. They are quite maneuverable, able to perform turns of various kinds while in the air. Flying snakes are capable of making 90-degree turns during glides and can travel 100 meters before landing. They make lateral slithering motions and flatten their body by extending their ribs to create a concave belly for lift.

Some flying lizards can glide for 60 meters while only losing 10 meters of height. Their ribs extend out to the sides and support the membranes that provide lift, which is unusual since it's more common for reptiles to use membranes on their limbs for this purpose. Flying squirrels glide between trees with the help of a membrane that stretches between the wrist and ankle, and use their tails for stabilization in glides as long as 200 meters.

Whether the issue is science or flying, names can be confusing. It would be especially hard to know what to make of something called a flying scientist.

Flying animals may not fly, but their names make them sound more exciting than if the word "gliding" were used instead. I still find it exciting that we could occasionally see flying squirrels in the back yard of our house in Wisconsin.

Birds Make Our Spirits Soar

December 3, 2013

After running as fast as he could to reach us, my son bent over in fatigue, paused to catch his breath and gasped, "Bat Falcon . . . perched!" His brother, uncle, grandparents, father, and I simultaneously leapt up from the breakfast table at the lodge, grabbed binoculars and ran. We followed him until we could see this spectacular bird at the top of a 90-foot-tall tree. We observed it until it left its perch, flying at a speed I can only describe as falcon-fast.

Our first task upon returning to the restaurant was apologizing to the owner for rudely interrupting our conversation with him by our abrupt departure. There had been some concern among us that we might even have shoved him. Thankfully, he was kind enough to deny it and assure us he was simply glad we were enjoying the birding opportunities on his property.

We have been amazed many times by the multitude of flying (and perched!) treasures in Costa Rica. While the Bat Falcon incident highlighted our fervor, sightings of truly magnificent birds happen repeatedly in this avian heaven. This is not surprising given that the country has more than 850 species of them in an area smaller than West Virginia, which has about 320.

Costa Rica contains so many species for several reasons: its location in the tropics on the land bridge between North and South America; its varied topography, climate, and soil types mean it has many different ecosystems; and it has coasts on two oceans. The birds are so plentiful that it's overwhelming at times.

We were enthralled by a Black Guan, and shortly after, a Resplendent Quetzal flew overhead. We were watching thousands of Black Vultures and Swainson's Hawks migrate in giant flocks when we were distracted by both a King Vulture and a trio of endangered Great Green Macaws. The mystical calls of a flock of Montezuma Oropendolas caught our attention just as we were recovering from the euphoria of seeing a pair of Mottled Owls.

We were thrilled to see a Rufous-tailed Jacamar, but found ourselves in a state of sensory overload when the same location offered up a White-fronted Nunbird, a Buff-rumped Warbler, a Crested Guan, a Spotted Antbird, a Black-headed Nightingale-thrush, and a Buff-throated Foliage-gleaner.

The hummingbirds are so plentiful here that we once paused for 20 minutes on a trail and saw Coppery-headed Emeralds, Violet-crowned Woodnymphs, Green Hermits, Green-crowned Brilliants, and Violet Sabrewings. Over the course of our 4 months here, we've seen 19 species of hummingbirds. (We've seen 3 during our 8 years in Flagstaff.)

We were elated to see various quintessential tropical birds: Chestnut-mandibled Toucans, Red-lored Parrots, Blue-crowned Motmots, Green Honeycreepers, Orange-bellied Trogons, Yellow-headed Caracaras, Sungrebes, Gray-headed Chachalacas, Northern Jaçanas, White-collared Manakins, and Scarlet Macaws. We've seen birds so great their names proclaim it: Great Kiskadees, Great Tinamous, Great Black Hawks, and Great Curassows.

It's a weird but high-quality problem to see enough stunning birds to become dizzy, but I'm getting used to it since it's such a common occurrence in Costa Rica.

The Bat Falcon incident is one of our family's favorite memories of the semester we spent in Costa Rica. Few experiences more accurately reflect who we are and what excites us. Update: We have now seen 8 species of hummingbirds in our 18 years in Flagstaff.

Sleeping Beauty

January 7, 2014

Right after our first son was born, my husband and I read an article claiming that a newborn's frequent wake-ups put parents on a sleep deprivation schedule that is similar to those used as a form of torture to induce psychosis. This explained a lot about our early parenting experience. Although we had expected to be tired, we didn't know that we would be psychotic!

It's tempting to think that only humans struggle with sleep deprivation, but this problem affects other animals, too. Disrupting the sleep of fruit flies causes cognitive difficulties. They take longer to learn new tasks and forget them more quickly. Sleep deprived reptiles become harder to wake when they do finally sleep, suggesting that they sleep more deeply to compensate for missed rest.

Fish go long periods of time without sleep during periods of intense parental care, and during spawning and migration. There's no evidence that it unhinges them, but it's hard to imagine that it provides any benefit to their sanity. At least they haven't suffered in experiments as rats have. Long-term total sleep prevention in laboratory rats can result in death.

Though we share the need for sleep with other animals, we should count ourselves lucky that we don't share the same level of risk associated with it. Few of us worry about predators or other dangers during slumber. In other species, that is not the case, and they employ many precautions that allow them to stay alive to wake up and face a new day (or night in the case of nocturnal animals).

Whales, dolphins, and seals are among the animals capable of unihemispheric sleep, in which only one side of the brain rests at a time. The half of the body controlled by that side of the brain remains immobile. Even when they surface to breathe, the sleeping half of the brain doesn't awaken.

Horses can sleep while standing by engaging mechanisms in their legs to stay upright even while the muscles relax. Since they are prey animals that must be ready to flee at any moment, remaining in a standing position is advantageous. This only works for light sleep. Horses must lie down for deep sleep, but they require only an hour or two of that each day. Horses in groups sleep better than solitary horses because they can take turns sleeping and watching out for predators.

Birds use both unihemispheric sleep and bracing mechanisms to keep themselves safe during sleep. Resting only half of the brain at a time results in birds that sleep with one eye open. To prevent falls, tree-dwelling birds "lock" onto a branch during sleep. Taut tendons resulting from the position of their knees cause their feet to clench the branch. Bats also lock onto a branch or other surface to allow their famously upside-down sleeping behavior.

Whether done upside or not, "Sleep is the best meditation," according to the Dalai Lama, but if you want confirmation of the wondrous beauty of sleep, just ask any new parents, psychotic or not.

Many parents miss the days when their children were little, but we are not among them. We enjoy them more and more as they get older, and that is in part because we are now sleeping enough.

The Significance of Sheep
January 21, 2014

Sheep mean different things to different people, in part because there are so many species and breeds of sheep. In North America, there are four kinds of wild sheep (Dall's, Stone's, Rocky Mountain bighorn and California bighorn). Worldwide, there are anywhere from 200 to more than 1,000 breeds of domestic sheep, depending on who does the classifying.

Sheep appear in the written word and many people are quick to think of the sacrificial lamb. Others remember that it was the sheep in Orwell's book *Animal Farm* who served as the propaganda machine for the new regime and whose initial chant of "Four legs good, two legs bad!" was changed so that later on the sheep endlessly bleated, "Four legs good, two legs better!"

Being a dog trainer and behaviorist, I think of sheep as expensive dog toys, meaning no disrespect to the sheep. I'm simply comparing the cost of tennis balls to maintaining a flock of sheep—I actually know people who have moved to farms and purchased sheep because their Border collies needed a job. Any mention of sheep also calls to mind the famous sheep Dolly, who was the first animal cloned by scientists.

There are many people whose main associations with sheep are nursery rhymes like "Mary had a little lamb" or "Little Bo Peep has lost her sheep and doesn't know where to find them."

The idea of finding sheep occupied my mind last summer as we traveled to Rocky Mountain National Park. Bighorn sheep was the species of animal we most wanted to see, and we were ecstatic to observe five of them grazing on a steep hillside. We were more excited to see them than any of our other mammal sightings, which included moose, coyote, deer, pika, marmot, elk, and pronghorn. Though they are rare, they are regularly spotted in this park, so we knew that this trip was our best chance to see them.

Once common, populations of bighorn sheep were decimated by hunting and diseases introduced by domestic sheep, such as scabies and pneumonia. By 1940, only small, isolated populations existed, though they were once so abundant throughout the west that there were two million individuals at the height of their success. These sheep, so distinctive in appearance because the males have large curved horns that can weigh as much as 30 pounds, are symbols of the Wild West and the ability to survive the harshness of this wilderness.

The populations of bighorn sheep have recovered significantly from their low point of 20,000 individuals, and continue to grow, which means that this type of sheep can also be considered a conservation success story. Thousands of them have been relocated to suitable habitat in the United States and Canada as part of a program to re-establish bighorn sheep throughout the still inhabitable parts of the range they occupied historically throughout western North America.

The populations are now so large that it's a real challenge to count them, even for insomniacs accustomed to the task.

Though bighorn sheep were previously considered endangered, they are doing better now. Their official conservation status is "least concern" and their populations are stable.

Can You Hear Me Now?
February 4, 2014

Ethologists (those of us who study what animals do and why they do it) like to refer to the four Fs of animal behavior: fighting, fleeing, feeding, and reproduction. When it comes to animal communication, these categories cover the main types of vocalizations. There are territorial calls ("Back off, buddy. I'm willing to fight to defend my turf"), alarm calls ("Danger! Run for your life!"), food-related calls ("Come check it out. I found the mother lode over here"), and courtship calls ("Hey, baby baby").

Less well known are "contact calls," which simply let animals know who else is around and function to maintain a connection, even when individuals are out of sight of one another. They are used to help animals stay in touch as they feed, fly, or engage in any of their daily activities. This allows groups to maintain their cohesiveness. The repeated honking of geese in flight is one example of a contact call.

Screaming is a common complaint for owners of parrots, but a little knowledge of contact calls can help prevent this problem behavior from developing. Pet parrots use contact calls with their owners, and it's important to respond to them consistently from the beginning.

In the absence of an answering call, many parrots will call repeatedly with increasing agitation, sometimes becoming nearly hysterical. With time, the screaming becomes a habit, and it's far harder to change this behavior than it is to prevent it in the first place.

The calls that newborn puppies make when they lose contact with their mother are sometimes mistakenly referred to as contact calls, but they are more appropriately called retrieval calls. They don't simply alert the mother to the presence of a puppy, but actually cause her to go find the source of the sound, and bring it back to join the rest of the litter.

Only puppies produce these calls, and only when they are lost. The calls are made continuously without variation in pitch, amplitude or duration until the puppy becomes utterly exhausted or is rescued. Though it's tempting to say that the puppy's calls prompt the mother to go retrieve her puppy, experiments have shown that she will retrieve any object making the sound, even if it's a speaker or other mechanical device.

Many species produce contact calls nearly continuously, and when they cease, the silence signals danger. It's a similar reaction to silence that makes parents say, "Uh-oh, it's *too* quiet," and rush to check on their kids because they experience a feeling of panic similar to when they hear the children scream in alarm.

Human humming may have originally functioned as a contact call, preventing that unsettling quiet. It has been proposed that many humans find full silence distressing because it signals danger to them. Perhaps that's why many people have the television or radio on constantly when alone and may also explain why we whistle or hum or talk to ourselves in solitude. At least we're not screaming into the void like parrots do!

Contact calls are usually short and inconspicuous, which may explain why they have been studied less than many other types of vocalizations.

Swifter, Higher, Stronger
February 18, 2014

Do you find yourself humming *Swan Lake* and considering taking up ballet? Making plans to visit the nearest museum with paintings by Kandinsky in its collection? Developing a strong urge to read Tolstoy, Dostoevsky, and Chekhov?

Interest in all things Russian is a natural consequence of having the 2014 Winter Olympics in Sochi, Russia. For me, that means a quest to learn more about the animals living there, and especially those that represent the Olympic motto "Citius, Altius, Fortius," which means "swifter, higher, stronger."

Swifter

Saiga antelope can run at 60 miles per hour, so they are certainly swift. They are also swiftly declining in numbers and critically endangered, having lost 90 percent of their population in a decade. This is one of the fastest declines of a large mammal observed in recent years.

Poaching is a problem because their horns are used in traditional medicine and habitat loss is also a challenge.

Saiga antelope have a distinctive appearance with a long nose that hangs over the mouth. The big nose is thought to filter out dust during the summer and warm the air during the winter. Throughout the breeding season, the unusual appearance is compounded in males. Their noses swell up still further and the tufts of fur below their eyes are covered in a sticky substance. In other bad news for the males during this time, they engage in violent fights that result in many deaths.

Higher

Snow leopards live high in the mountains in the far northern parts of Central Asia at altitudes ranging from about 10,000 to 17,000 feet. Their large paws are covered with fur and act like snowshoes, preventing them from sinking into the thick snow so common where they live.

They are not actually Kung Fu experts, whatever Tai Lung from the movie *Kung Fu Panda* may try to make you believe, but they have their own style of athletic abilities. They spring at their prey of wild sheep, wild goats, birds, and rodents, leaping up to 30 feet, which is six times their own body length.

There are only about 5,000 of these animals remaining in the wild, which combined with their elusive behavior, camouflage, and rugged mountain habitat makes sightings of them very rare.

Stronger
Polar bears are strong enough to kill animals with one swipe of a massive paw, but their bite strength of 1,200 pounds per square inch is more impressive. That bite strength is twice as powerful as a great white shark.

These animals have strong senses of hearing and smell, too. They listen for prey such as seals and fish, which they can hear through up to 3 feet of ice. They also eat birds and berries, along with making meals out of beached dead whales, which they can smell from several miles away. Polar bears have skills that allow them to do far more than appear in soda commercials and make people think about climate change.

Just like the world's Olympians, these Russian animals are capable of amazing feats.

Conservation groups in Russia target preserving habitat, especially the forest and tundra in the far eastern areas of the country. Efforts are helping save the species mentioned above as well as Amur tigers, fish owls, and Kamchatka brown bears.

Sea Devil Romance
March 4, 2014

Even though Valentine's Day is now in our rearview mirror, the fact that Flagstaff can be a tough place to meet someone special remains on many people's minds. Yet, it's definitely not as hard to meet a partner in this town as in some locales. Consider the challenges facing those living in the desolate waters of the deep ocean and take heart.

In that vast habitat, a fish can swim for days without encountering another fish of any kind. That makes the likelihood of coming across a member of your own species of the opposite sex somewhat low. The sea devils are a group of anglerfishes whose behavior and anatomy have both been influenced by that rarity of encounters, resulting in the evolution of a curious mating system.

The sea devil's method of reproduction ensures that when a female is ready to spawn, sperm are available to fertilize her eggs. Free-living males have one purpose—to find a female. They need to do that because they are unable to survive very long on their own. The males have large olfactory organs and an excellent sense of smell, and scientists believe that they find mates by sensing specific pheromones released by females.

Once a male accomplishes his life's goal of finding a female, he releases an enzyme capable of digesting his own mouth as well as the skin of the female. They become permanently attached and their blood vessels fuse. Only then do they both mature and develop the ability to sexually reproduce. Females without an attached male never reproduce because they do not mature, and males die if they fail to find a female within the first few months of life.

After they fuse together, the male becomes larger, continually receiving oxygen and nutrients from the female through their shared circulatory system. Much of his body degenerates, but he does grow large reproductive organs that produce sperm. A female may have as many as eight males attached to her, almost always on her belly and upside down, facing the same way as the female, suggesting they approached her from behind prior to attachment.

Females are tens or even thousands of times larger than males. In one species, the largest female ever measured was 500,000 times larger than the smallest male ever measured. If that were true of people, it would mean that based on the heaviest woman (1036 pounds), the smallest man would weigh much less than an ounce. Male and female anglerfish are so different in appearance and size that scientists initially failed to realize that they were members of the same species.

Researchers originally thought that only female sea devils had ever been caught, which was puzzling. It turns out that plenty of males had also been caught, but they were attached to the females' bodies and so small that they were incorrectly but understandably identified as parasites.

In fact, this unusual method of reproduction is called "parasitic reproduction." The name surely makes clear that although the Flagstaff singles scene may not be perfect, things could be worse.

The anatomical fusion of different individuals would not be possible in most species because their immune systems would reject and attack each other. Anglerfishes have very different immune systems than

those of other species, making them interesting for the study of immune function and possibly important in the development of new ways to help people with immune disorders.

Temperatures Troubling to Turtles, Too

April 8, 2014

All I have to say about the crazy weather across the U.S. this winter is that it's a good thing we're not all turtles.

If we were, the extreme cold in the north and east and the unseasonably warm temperatures here in the southwest would make the balance of new baby boys and girls an out-of-whack nightmare everywhere. We'd likely have an excess of male babies born where winter hit hardest and nothing but female babies born where milder temperatures prevailed. It's hard to contemplate how online dating sites would cope with that a few decades from now.

Many turtles have temperature-dependent sex determination, meaning that the temperature of the environment during development determines whether eggs develop into males or females. In Pattern 1, if environmental conditions are hotter than some threshold temperature, females will develop, but if they are lower, males will hatch from those eggs. Though less common, there are species that follow Pattern 2, in which males are produced at intermediate temperatures but females develop in conditions that are more extreme—either hotter or colder.

This only seems weird to us because it's a completely different system than what occurs in mammals, including humans, where sex determination is genetic. Females have two X chromosomes while males have one X chromosome and one Y chromosome. During reproduction, each parent gives a copy of one of these chromosomes to every offspring. Moms always give an X, but dads give either a Y, which produces a son, or an X, which produces a daughter. That's an oversimplification of the variation found in nature because there are also humans who have two Xs and a Y (Klinefelter Syndrome males), one X and two Ys (XYY Syndrome males), or even only a single X (Turner Syndrome females).

Birds are similar to mammals in that sex determination is genetic, but it's the males who have two matching chromosomes. To distinguish this system from the XY system of mammals, in the avian world, it's called the ZW system. Males have two Z chromosomes, while females have one Z chromosome and one W chromosome. So, in birds, it's the females whose genetic contribution determines the offspring's sex. That's just a minor difference compared to the genetic mechanism of sex determination in most social insects.

In bees, ants, and wasps, if a reproductive female lays an egg without fertilizing it using the sperm she has stored since mating with a male, that egg will develop into a male. However, if she releases sperm to fertilize the egg, it will develop into a female. There is unequivocal evidence that females choose whether or not to produce males or females based on current environmental and social conditions.

We all have reason to complain about this past winter, whether we have suffered through relentless cold and snow or missed out on winter entirely and now dread the upcoming fire season. On the bright side, at least we still have high hopes of a proper mix of newborn girls and boys throughout the nation.

As climate change worsens, so does the issue of sex ratios in turtles. Many other species receive more attention and concern for how climate change affects them, but turtles have a particularly big challenge as extreme weather events become more common.

Island Biogeography
April 22, 2014

In addition to animals, I love islands, simple explanations of complex phenomena, and math. Math can be beautiful, but before I lose too many of you completely, let's get back to islands.

In the 1960s, scientists Robert MacArthur and E.O. Wilson launched the field of island biogeography by developing a theory, elegant in its simplicity, to predict the species richness of islands. Their theory postulated that the number of species on an island is a function of both colonization and extinction and can be calculated mathematically. Islands gain diversity when new species arrive and lose diversity when a species goes extinct.

Two main factors determine how many new species arrive on an island. One of them is the size of the island and the other is the distance of the island from the mainland that is the source of new species.

Most colonizations of islands are highly unlikely random events. Perhaps a bird is blown badly off course by a storm but rather than falling into the ocean exhausted, actually survives to land on the island. Or maybe a small piece of land on the mainland breaks off and floats to the island, arriving before everything on board this raft dies. Such rafts are most likely to contain plants and insects, but occasionally an unlucky passenger such as a rodent or a snake may be on it. Seeds may drift or blow to the island.

The bigger and closer the island is to the mainland, the more probable that such accidental travelers will make it to the island alive and reproduce there, starting a population of a species not previously found on the island.

To visualize these principles, imagine that you are the source of new species, represented by balls of crumpled paper. Your task is to throw the paper balls toward the "islands," which are four pieces of paper on the floor. Two of the paper islands are close (10 feet away) and two are far (25 feet away). The close and the far island pairs each consist of an island that is an 8½ by 11 sheet of paper and one that is a 2-by-2 sticky note.

The theory of island biogeography predicts that the most species (balls of paper) will land on the close island that is big and the fewest will land on the small island that is far, which makes intuitive sense. Your odds of hitting a small target far away are lower than a large target that is close. Most migrants are accidental travelers whose prospects of arriving at an island are subject to chance, just as those paper balls are.

On the extinction side, the bigger the island is, the more likely it is to have habitat that suits a new migrant. Additionally, an established population is less likely to go extinct on a bigger island, because bigger islands can support larger populations, which are more resistant to extinction than smaller populations.

The theory of island biogeography encompasses an appreciation of many things including islands and yes, math.

When I taught high school students this concept, I had the students throw paper ball "species" toward imaginary paper "islands" of different sizes and at different distances and we recorded how many species landed on each island. The results matched the predictions made by the theory of island biogeography.

Elephant Moms—And All Moms—Deserve Applause

May 6, 2014

They say elephants never forget, but I hope that's not true. Perfect recall means that female elephants will always remember their pregnancies, and if I had been forced to carry my unborn babies for nearly 2 years, I would probably welcome a little amnesia. We human moms typically carry our babies for 266 days, which is a few days less than a full 9 months. (FYI, since you may be clobbered for pointing out the "less than" part of this quantity to most women, it's safer to stick with the party line of 9 months.)

In contrast, African elephants have a gestation period of 645 days, or 21½ months. That's a long time to grow another individual, sharing your own personal oxygen, fluids, nutrients, and energy. In fact, elephants have longer pregnancies than any other mammals.

There are general patterns to the length of gestation periods, defined as the time between fertilization and birth. The amount of time animals spend developing inside their mothers before being born depends on a couple of factors. Generally speaking, the bigger an animal, the longer its gestation period.

Mice and rabbits are pregnant for 21 and 31 days respectively. Dogs carry their babies for approximately 63 days, cheetahs for 93, lions for 108, chimpanzees for 227, and camels for 400. The largest rhinoceros species has a gestation period of 540 days, which is 105 fewer days than the aforementioned elephants.

Another factor that influences the gestation period of animals is how developed they are at birth. If the young are immature at birth, the gestation time will be shorter, as is the case in marsupial mammals. These are the mammals whose young are born poorly developed and continue their maturation in their mom's pouch. The short-nosed bandicoot, which gives birth about 12.5 days after fertilization, has the shortest known gestation time. (You know it's short when fractions of days are involved.) Red kangaroos have the longest gestation period among marsupial mammals at about 33 days.

Animals whose lives depend on a high level of development at birth have longer gestation periods than their size predicts. For example, horses must be able to walk and run within hours of birth to avoid predators and have longer gestation periods (336 days) than predicted based on their size alone. Based just on size, we would expect horse pregnancies to last somewhere between the 220 days of the grizzly bear, which is smaller, and the 285 days of the American bison, which is bigger.

Luckily, I think it's impossible that elephants remember everything about their pregnancies. In my experience, there is a certain amount of forgetfulness that comes along with "sharing a brain" for so many months.

I hope all moms have a happy Mother's Day, whether you carried your baby for (approximately) 9 months, traveled around the world to adopt, did so right here, or found your way to a child in any other way.

However we become moms is worth it, and that includes the elephant way.

Elephants can give birth about every 4 years, which means that while they are reproducing, they are pregnant nearly half the time. For an elephant, who may have 4–12 babies, that means roughly 7–21 years of pregnancy.

The Sting of Colony Collapse Disorder
May 20, 2014

Besides walking to school through snowdrifts uphill both ways, we adults differ from the youth in our community in another way. We regularly ran barefoot through fields of clover (or on the grass at the local park anyway) and became afraid of bees after being stung by them. Younger people may be deprived of this experience. Honey bee populations are declining worldwide due to the little-understood phenomenon called Colony Collapse Disorder, and that is very bad news.

Perhaps you consider this good news and wonder why we shouldn't respond to the decline of honey bees with high-5s all around. The answer is that honey bees are essential for our health—economic and otherwise. The pollination services of honey bees are worth 15 billion (yes, billion with a "b") dollars to the agricultural economy in the United States, and at least a third of our food supply depends on them.

Apple orchards need one healthy colony of bees per acre for adequate pollination, and almonds are completely dependent on bees for all of their pollination. Last year, many almond growers in California were unable to get enough colonies to pollinate their crops, even when willing to pay far more than usual. Each year, 1.5 million hives are temporarily relocated to California for a few days or a week for their pollination services. Hives-for-hire is big business, and without healthy bees, food supplies will suffer because a huge number of fruit, vegetable, and nut crops depend on bees for pollination. Inadequate production means higher food prices.

With the dependence of our food supply on bees, we are in real trouble given the massive mortality rate of bee colonies in the last decade. In recent years, many beekeepers have lost a third of their hives each year, and some beekeepers have lost all of their hives to Colony Collapse Disorder.

The sign of CCD is the disappearance of large numbers of worker bees. Bees leave the hive to forage, but so many fail to return that the colony cannot sustain itself and dies, despite a healthy queen and growing larvae in the hive.

The causes of Colony Collapse Disorder are not clear, despite an abundance of possibilities. Multiple factors may be at work in this massive decline of honey bees.

1. Environmental stress: Pesticides and insecticides are prime suspects, and that includes chemicals present

in polluted water sources. In addition, pollen and nectar scarcity because of the conversion of so much grassland to corn and soy fields may stress bees and make them vulnerable.

2. Pathogens: Although no single pathogen is directly correlated with colonies affected by CCD, higher amounts of viruses, bacteria, and fungi are found in such colonies.

3. Parasites: Varroa mites are often present in affected colonies, though it is unknown whether the mites themselves or the viruses they carry are a factor.

Because of Colony Collapse Disorder, we may not risk being stung on the foot as often anymore, but the food shortages and hit to our wallets will "sting" far more in the long run.

Pollination shortages are becoming more serious as both wild bees and managed honey bees continue to decline in numbers. The effects on food production and food costs are expected to keep getting worse.

Animal Fathers
June 3, 2014

If you're lucky enough to have a Dad who has watched over you, played with you, provided you with anything you needed ranging from food to encouragement, or protected you from harm, then you have benefited from what ethologists call paternal care. Paternal care is investment by a male in offspring, and though it's not the norm in the animal world, it's certainly not unique to humans. In fact, it occurs in a number of taxa, including insects, amphibians, and fish.

Parental care that is performed exclusively by males is rare in the insect world. Of the more than one million described species, no more than about 150 have paternal care only, and almost all of them are in the family Belostomatidae.

Belostomatids, or giant water bugs as they are commonly called, are well known for the care that fathers give to their offspring. In these insects, only the fathers provide any parental care. The females are done with their parenting work when they have mated with the father and laid up to 150 eggs on his back.

The father takes it from there, caring for the offspring in several ways. He regularly rises to the surface so that the eggs can receive more oxygen than they can while in the water. He also moves around in the water performing a behavior that looks remarkably like push-ups, agitating the water near his eggs, also in order to increase their levels of oxygenation. Eggs that are deposited anywhere but on the back of a male will fail to hatch, but almost all of the eggs that receive this paternal care do hatch.

As with the giant water bugs, female Darwin's frogs lay eggs and then take off, leaving the male to care for the young. Darwin's frog is a species discovered in Chile by the famous scientist himself in the 1830s, but it was decades later in the 1870s that their unusual form of paternal care was described. The male guards the eggs for about 3 weeks until they are just about to hatch, at which point he swallows them. The young stay in his vocal sac for the rest of their development, including the tadpole stage. About 2 months later, they emerge from his mouth as fully formed frogs.

Seahorses are another group in which only the males care for the young. The female lays about 1,500 eggs, which develop inside a pouch on the male's tail. He controls the temperature, salinity, and oxygen levels in the pouch to provide the best environmental conditions for the development of the young. He carries the eggs for up to 45 days and when the tiny but fully developed baby seahorses emerge, his job is done. He is then ready to mate again and start the process all over again.

Humans have a high frequency of paternal care, and as a society, we all benefit from what dads do. Happy Father's Day to all fathers, stepfathers, and every other type of paternal caregiver out there!

Humans, like many primates, exhibit a lot of paternal care.

Handling the Desert's Dryness

June 17, 2014

It's that time of year again when my skin makes the average alligator seem soft and smooth by comparison. I'm currently going through as much lip balm as shampoo, and no matter how much I drink, proper hydration remains an elusive goal. The question that has crossed my mind a number of times in the 9 years since moving to Arizona is: "How does anyone survive here?" For most people, the answer is that it's done with difficulty, vigilance, and constant pharmacological intervention.

For many animals, the answer is that millions of years of evolution have led to adaptations to such an arid climate. Of course, in the high mountain desert that I love so dearly (even at this time of year), the harshness we face is mild compared to the extremes of the Sonoran Desert in the southern part of our state. While I may ask in jest how anybody survives up here, the question has no trace of humor when directed toward the animals that thrive in hotter, drier conditions. And yet, many of them do quite well because of the adaptations that make them well suited to life in the desert.

One animal that is well known for its exceptional ability to handle dry conditions is the kangaroo rat. Kangaroo rats are able to survive in the desert without ever drinking water. They are able to do this because of their many anatomical, physiological, and behavioral adaptations to the arid conditions.

Though they do sweat, water loss due to this process is minimized because they only have sweat glands on their feet. They do not lose water through panting, which is unusual among mammals. Kangaroo rats lose less water through elimination than other mammals because their efficient kidneys are able to maximize reabsorption of water and produce highly concentrated urine.

Kangaroo rats spend most daylight hours in burrows where it is cooler and they do not have to sweat or salivate to cool off, which would cost them precious water. Burrows have higher moisture levels than the outside air, which helps them minimize the amount of water they lose from breathing. They emerge after the sun has set.

They eat seeds, which are high in carbohydrates and therefore produce water when they are metabolized. The amount of water produced is small, but kangaroo rats are so efficient in their use of water that they're able to survive without additional water sources. They choose seeds with the highest water content, searching over large areas to collect the seeds that yield the most water.

Though kangaroo rats are able to survive without drinking, the amount of water in the environment does affect them. Their reproduction is tied to rainfall. Many measures of reproductive success vary with the amount of rain, including embryo number, embryo size, testis size, sperm production, and ovary size.

Kangaroo rats are water conservation experts, and though they thrive in dry climates, I have no idea how they cope with the unavailability of Chapstick™ in the desert.

As a kid, I once captured kangaroo rat footprints on an index card coated with smoke from a kerosene lantern and left out overnight. These animals have had a place in my heart ever since.

Surprisingly Related
July 8, 2014

Elephants would not be in most people's top 100 answers if they were asked what animals are most closely related to the hyrax. The majority of guesses about where they fit in taxonomically would have them linked with rabbits, guinea pigs, or squirrels, though some people might notice that hyraxes look like chubby meerkats. Yet, they are more closely related to the elephant than to these smaller animals, which is a bit surprising because at first glance (as well as at second and third glances) they seem nothing alike. Hyraxes, which live in Africa and the Middle East, weigh around 10 pounds. They have thick gray or brown fur and small ears.

If we based our guesses on superficial morphology—the overall look of the animals—we would also make very different claims about who elephants are related to. Elephants resemble the hippopotamus and the rhinoceros, which are also massive animals, and though it's a reasonable guess that these are the elephant's closest relatives, it's not correct.

Stunning variation exists between relatives, including close relations. It is often incomprehensible how such differences can exist, but they are far from unusual. Most of us have a relative who we assume nobody will think is part of our family because we are so different, and their presence can seem curious. Or, as my own mother explained it to me as a child, "Every family has an Uncle Norman."

Even closely related animals can be very different in appearance. As they evolve and adapt to their respective environments, morphological changes make the animals appear unalike in many ways even as they maintain shared traits that they inherited from their common ancestor. Although the basics of size, shape, and fur type don't match, hyraxes and elephants have many similarities.

The elongated teeth of both of these animals evolved from incisors into tusks. It's the canines that are the origin of tusks in most other mammals. Hyraxes and elephants both have toenails, which are flattened rather than long and curved like claws. Both have pads on the bottom of their feet and the same pattern of toes with four on each front foot and three on each back foot. This may seem trivial, but the number of digits is an important trait for classifying animals and analyzing relationships between groups of animals.

The testes of hyraxes and elephants do not descend into a scrotum, but are located in the body cavity, which is highly unusual among mammals. Another trait unusual among mammals and shared by elephants and hyraxes is the location of mammary glands between their front legs.

Both animals also have excellent hearing and superb memories. Oh, and they have one other thing in common. The third group of animals that make up this group of close cousins is the sea cows (dugongs and manatees), which, despite appearances, are not closely related to the walrus or any other marine mammals. Hyraxes, elephants, and sea cows make up one of the weirdest groups of cousins in the animal world.

The idea of relatives looking very different is something I've thought of often since my sister Marla, who looks quite different from me, once said, "We're not related. We're only sisters."

A Worm of Legend
August 5, 2014

While in Florida recently, we saw diverse animals such as Sandhill Cranes, West African manatees, alligators, leopard frogs, and bluegill. Naturally, I want to write about the most fascinating animal we saw during our week in the swampy southeast, and that animal is . . . the Gordian worm. We saw it in a glass on our host's kitchen island. He's a biologist who is interested in all animals, and he found it in his swimming pool that morning. Naturally he was eager for us to see it.

This worm's name refers to the Gordian knot of legend. According to mythology, Gordius was a peasant who became king. To honor the occasion, his oxcart was dedicated to a god associated with Zeus by tying it to a pole with a complicated knot that was allegedly impossible to unravel.

Despite this, an oracle made a prophecy that the future king of Asia would untie it. Though many people came to town to attempt to do so, it remained fastened until Alexander III of Macedon came to visit. Frustrated at his inability to find the ends of the knot, he slashed through it with his sword instead, thus undoing it. He fulfilled the prophecy by becoming the King of Asia, a feat that explains why we know him better as Alexander the Great.

"Cutting the Gordian knot" is an expression that means to solve a difficult, complex or seemingly impossible problem by taking bold action. The Gordian worm takes its name from the legendary Gordian knot and refers to the fact that these 2- to 4-foot long worms regularly tie themselves in ball-shaped knots of baffling complexity, either alone or in clumps of many other worms.

Adult Gordian worms live throughout the winter in water such as ponds, streams, puddles, or pools. They mate in the spring and the female then lays her eggs in the water. In 3–4 weeks, the larvae hatch and must soon enter a host to survive. The hosts of Gordian worms are often long-lived insects such as dragonflies, water beetles, praying mantises, grasshoppers, and crickets. These hosts ingest the larvae, which develop into adults within them, feeding on their insides.

Approximately 3 months after a Gordian worm parasitizes its host, that host will seek out water. This behavior occurs as a direct result of the worm's presence within its body, as the worm essentially has the ability to "instruct" the host to go jump in a lake, even though that means that they will likely drown. When the host enters the water, the fully mature worm emerges. Adult

Gordian worms don't eat, though they can live for months, overwintering until they mate in the spring.

Although their name reflects the tendency of these worms to form knots with their bodies, it could also refer to the fact that they have a complicated life cycle, which solves the problem, faced by all species, of how to reproduce and disperse, but in an unexpected (dare I say bold?) manner.

Gordian worms are also called horsehair worms because of an old belief that they were horsehairs that fell into water and came to life or cabbage hair-worms because they are found in water that collects on cabbage leaves.

In Defense of Poop
September 9, 2014

If you are one of those individuals who only ever defecates in the bathroom (or in the woods on special camping occasions), then you are missing out on some practical protective uses of excrement. Other species of animals offer inspiration in the creative ways that they defend themselves with this readily available substance.

Many primates hurl their feces at potential predators. Anyone bothering or harassing them is at risk of being hit by poop because many apes and monkeys throw it when they feel threatened, angry, or irritated.

My life list of monkey species that have thrown poop at me in the wild contains only two entries: white-faced capuchins and wedge-capped capuchins. Additionally, I've seen mantled howler monkeys do this to people near me, but unluckily, I wasn't close enough to be part of their target. Darn. It turns out that a variety of monkeys and apes also exhibit this behavior when bored, which explains why people so commonly scream near primate exhibits at zoos.

Researchers studying feces throwing by captive chimpanzees made the startling discovery that the chimps who threw poop accurately and threw it often were more intelligent and had higher brain function than chimps who threw it less often or hit their targets infrequently. That's counterintuitive, to say the least.

According to brain scans, the accurate and frequent throwers had greater development of both their motor cortex and Broca's area, which is a part of the brain that's important for human speech. The good throwers were better communicators within their social group, though they did not have superior physical skills. The act of throwing may be an important step in the development of speech in humans, and throwing objects may have served as a form of communication before being used as a defensive strategy.

It's not only primates who fling feces about. The Fieldfare is a type of thrush that also uses it against enemies. This bird, which is related to our American Robin, is about 10 inches long, very social, and lives in many areas of Europe and Asia. When avian intruders such as crows or owls approach Fieldfare nesting sites, the entire flock works together to protect their young by flying over and defecating on potential predators. The wetness and the weight of the feces are both damaging to predators. The extra weight makes flight impossible and the wetness interferes with the ability to thermoregulate, sometimes causing death from cold. The Fieldfare's scientific name, *Turdus pilaris*, seems to allude to this defense-by-feces behavior, but *turdus* is actually Latin for "thrush."

It's not just predators that can be rendered harmless with the judicious use of feces. Bacteria are also vulnerable to it. Vultures can eat rotting meat without getting sick because the chemicals in their digestive system kill dangerous bacteria. So, when they defecate on their own legs, they are not only cooling themselves down through the effects of evaporative cooling, they may also be giving themselves an antibacterial wash of sorts.

Poop—who'd have thought it was so serviceable?

I feel bad knowing that many people like to read the newspaper as they eat breakfast. Fun facts are not always appetizing.

Fun with Names
September 23, 2014

It's a rare name that doesn't open one up to being laughed at on occasion. For example, I love my name and consider myself fortunate to have it, but as a child, people would occasionally sing, "London Bridge is falling down" when I took a spill. Years later, I'm still not fond of the song. I have it pretty easy compared to most of my human friends and have no cause for complaint when my troubles are weighed against those of many animals.

Not all species are named elegantly or fairly. The Great Tit (related to our chickadees) and the Satanic Nightjar are just two of the many species whose names often incite laughter. The monikers star-nosed mole, Venezuelan poodle moth, blobfish, and Greater Pewee result in a chuckle (or at least a raised eyebrow).

Some names certainly fail to conjure up images of respect. The pleasing fungus beetle, the Dumbo octopus, the whirligig beetle, and the eyelash viper all sound less than dignified. The earwig, naked mole rat, aye-aye, and red-lipped batfish have similar challenges. The vampire moth, robber fly, and assassin bug sound like they're trying to be threatening, but couldn't possibly be. Some names just sound silly, including Mustache Puffbird, poison dart frog, handsome fungus beetle, backswimmer, violin beetle, and Star Wars wasp. (If you like that last one, you'll be pleased to know there's a genus of mites named *Darthvaderum*.)

It's not just common names that show more creativity than kindness. Scientific names illustrate that even in Latin, a joke is still possible, and it's often on purpose. The genus of parasitoid wasps *Heerz* calls out to scientists, who named one species *Heerz lukenatcha* and another *Heerz tooya*. *Pieza cake*, *Pieza pi*, and *Pieza deresistans* are small flies whose experts have a sense of humor.

It could be argued that the idea of Latin names was taken too far with *Vini vidivici* and *Ytu brutus*. The former is a slight variation on the Latin phrase "Veni, vidi, vici" which means "I came, I saw, I conquered." The common name of this species is Conquered Lorikeet, a South Pacific bird that went extinct about 1,000 years ago. The latter is the Spanish translation (¿Y tu, Brutus?) of what are commonly considered Julius Caesar's last words (Et tu, Brute?) It's a big name for a small Brazilian water beetle.

Spanish shows up again in the moth *La cerveza*, which means "beer." Any genus named "*La*" poses an invitation to naming humor, so it should come as no surprise that members include *La paloma*, which means dove, and *La cucaracha*, which means cockroach.

Sometimes the Latin names aren't actually Latin, as in the snail *Ba humbugi*, and the sea snails in the genus *Ittibittium* that are, in fact, smaller than the related snails in the genus *Bittium*. The unicellular organism *Kamera lens* seems to have received the name just for fun.

These names and many others provide evidence that a sense of playfulness is alive and well among biologists.

I liked my name better before "Karen" came to mean an obnoxious, entitled, angry, racist, usually blond, white woman who wields her privilege as a weapon to hurt others. A term was needed for that type of character, but I make sure not to be her even if I have the name.

Large AND Loud
October 7, 2014

It's common knowledge that the blue whale is the largest animal that ever lived, including all of the dinosaurs. It's less well known that this species is also the loudest animal in the world with calls that can reach 188 decibels. To put that in perspective, human yelling is about 70 decibels, and jet planes come in at about 140 decibels. Sounds over 120 decibels typically hurt people's ears, and eardrums can rupture when exposed to 160 decibels.

Blue whales use their vocalizations to communicate with other individuals about their location and to find and attract mates. Whales produce sounds in the larynx, although they do not have vocal cords in their larynxes. They also produce calls by moving air between their nasal sacs. Larger males can take in more air and hold the notes longer, making them more attractive to females.

Blue whales have dialects, based on where they live. So, blue whales from near Antarctica sound different than those off the coast of Chile, and those from the eastern part of the Pacific don't sound the same as those from the Western Pacific.

No matter the accent, blue whale calls can travel long distances. These low frequency rumblings can carry across more than a thousand miles of ocean because water is a great conductor of sound, a fact that has been known for hundreds of years. Leonardo da Vinci noted in the 15th century that if you put a tube in the water and listened through the end protruding from the surface, you could hear far away ships moving through the water.

Although sound travels well through water, light does not. Light is better suited to moving through air than water, but sounds travel 4–5 times faster through water than air, and the low-frequency sounds of whales travel significantly further, too.

It's no wonder that many sea creatures have evolved to use sound as a navigational tool, to find food, and to communicate. Unfortunately, sound is less effective now than in the past. When da Vinci talked about hearing distant ships, he was referring to sailing ships, which are far quieter than the propellers and engines that have made our oceans so much noisier than in his day.

The noise in today's oceans includes far more than wildlife, waves, and the sounds from natural events like undersea earthquakes, volcanic eruptions, and lightning strikes. Now, there are the sounds of vessels from cruise ships to supertankers, sonar from the military, and explosions from efforts to find oil and gas as well as to drill for them. Noise pollution poses a major problem for blue whales. Due to background noise, they can now only hear calls from roughly 100 miles away rather than from more than a thousand miles away.

Noise interferes with their communication system, including mate attraction, posing a threat to attempts to restore the population of blue whales to sustainable levels. Being loud and large does not protect them from the system-jamming effects of today's noisy oceans.

The ocean seems so peaceful to me when I am scuba diving or snorkeling, but to blue whales (and many other marine organisms), it is a raucous, earsplitting environment.

Black Cats: Lucky or Unlucky?
October 21, 2014

A black cat crossed my path this morning, which I took as a bad omen, naturally assuming that it meant I would not make my deadline for this column. (As a writer, it's hard to imagine anything more dreadful than a deadline passing me by.) Many people confronted by black cats presumably imagine even worse bad luck, but that feeling is hardly universal. Views of black cats have changed over time, and vary around the world.

In ancient Egypt, all cats were revered, including black ones. Cats were considered so valuable that laws protected them from being killed or even injured. Families mourned the death of their cats, embalming and burying them with great ceremony and expense. Cat cemeteries from this era contain many mummified black cats.

Black cats were regarded with suspicion in England during the Middle Ages when people thought that witches turned into cats at night. One story is that a father and son saw an injured black cat limp into the house of a woman suspected of being a witch. The next morning, that woman was limping, bruised, and had her arm bandaged. This convinced the townspeople that she was a witch by day and transformed into a cat each night.

Many societies tried to eradicate black cats during the latter part of the Middle Ages. Suspected witches and their cats were burned at the stake in England as well as in France. It's remarkable that black cats persisted in the face of these murderous rampages.

Their continued existence indicates that black cats have survival advantages sufficient enough to counteract longstanding attempts to eliminate them. Black fur has evolved in almost a dozen feline species, suggesting that it has survival advantages. Scientists have found at least one. The mutation that causes black fur may also provide protection from viral infection. The relevant gene is called MC1R, and it's in a family of genes involved in transmembrane receptors. These receptors provide entryway into cells that bacteria and viruses use to infect the cells. The mutated form (that also gives cats black fur) may prevent such infection.

It makes more sense, then, for black cats to be considered favorable omens, which they actually are in many cultures. In Australia, Japan, and Scotland, black cats are considered lucky. In the South of France, good luck is said to come to those who feed black cats and treat them well.

In Yorkshire, England, many believe that a black cat in the house will keep a fisherman safe until his return. The association of good fortune with black cats is not new to English sailors. Long ago, they thought that a black cat kept happy in their homes would guarantee them good weather at sea. A black cat in the theater on opening night is good luck in the United States and in Europe, and many English brides were given a black cat as a wedding gift to bring good luck.

Happy Halloween to all who love (and all who fear) black cats!

It's been bad luck to be a black cat in some areas of the world recently, due to the false idea that consuming them is a cure for COVID-19.

Who Bit Me?

November 18, 2014

He bit me!" is a common response to mosquitoes, but it contains a factual error since only female mosquitoes bite. Similarly, stinging bees and wasps are all females while the males are harmless. These are a few basic insect facts that aren't widely known. There are many more.

Insects are animals. Phrases like, "There are so many animals, birds, and insects in the tropical rain forest," make no sense because insects (as well as birds!) are most definitely animals.

In keeping with the theme of which organisms fit into which categories, spiders are not insects. Both insects and spiders are arthropods, along with lobsters, crabs, mites, ticks, scorpions, centipedes, and millipedes. All arthropods have an exoskeleton and jointed appendages, but insects have six legs and three body parts, while spiders have eight legs and two body parts.

On the subject of anatomy, insect legs attach to the thorax, which is the middle body section. Legs are often depicted on the posterior part of the body, which is the abdomen, and that is biologically inaccurate. Also, only adults have wings. Juvenile insects have not yet developed them.

Butterflies, bees, beetles, dragonflies, grasshoppers, cicadas, stink bugs, mayflies, and most other adult insects have four wings. Flies are an exception, and the scientific name for flies is *Diptera*, from the Greek "di" meaning two and "ptera" meaning wings. Whenever someone wears a honey bee costume with a single pair of wings, I see a person dressed as a hover fly, not a bee. Hover flies are bee mimics that typically have black and yellow stripes, but like other flies, they have only one pair of wings.

Bees do not gather honey from flowers. They collect nectar from flowers and use it to make honey.

Many insects look completely different as juveniles than as adults. It's common knowledge that caterpillars turn into moths or butterflies, but many other insects have similar transformations. As insects age, they molt (shed their exoskeleton), grow some more, and molt again. After 1 to 17 molts (3 to 5 is typical), they become adults, having reached their final developmental stage and size. Once they are adults, they stop growing. This means that little ants are not baby ants. A little ant is a little adult and a big ant is a big adult.

Insects don't see thousands of identical images with their compound eyes. The visual information is integrated to form an identifiable image, just as the input from our two eyes does not (usually!) result in double vision.

"Bug" is not a synonym for "insect" because only some insects are bugs. The term "bugs" refers only to a small number of insects such as cicadas, stink bugs, leafhoppers, and aphids.

It's well known that bees die when they sting you and can therefore only sting you once. Less well known (and arguably equally important) is that wasps and ants can sting you over and over. Mosquitoes can get you multiple times, too, making it perfectly correct to say, "She bit me again!"

Having studied wasps, I think it's pretty cool that they can sting multiple times, but I dislike very much that same multiple-strike capability in mosquitoes.

Life on the Farm

December 9, 2014

"Wow, that is quite an odor," a friend gasped when she came over to walk dogs on the pig farm where I lived.

Luckily, the wind usually took the smell away from the farmhouse I was renting, but today was an exception, and it was not so fresh by my home. For anyone who has not inhaled deeply around air infused with pig excrement, let me assure you that it's nothing like the normal earthy farm smell created by cows, horses, or sheep. Pig farms have a foul aroma that is really something special both in terms of its bouquet and its power.

The fact that I loved living in rural Wisconsin enough to put up with the smell of a pig farm showed how far I'd come since growing up in the suburbs of Los Angeles. My love of animals led to the decision to live there.

The 2 acres devoted to the house, lawn, and gardens plus the few with the barns were a small part of the 150-acre farm. Most was dedicated to growing corn or soybeans, depending on the year. (The town I lived in was called Black Earth, which has soil every bit as agriculturally productive as you'd expect.) A few dozen acres were wooded, which is what made life ideal for my dog.

Having open space to let my dog Bugsy run was so important to me that I ignored all other considerations, including the smell. I spent my days working as a canine behaviorist specializing in aggression and other serious behavior problems, my evenings training dogs in group classes, and weekends experimenting with life on the farm. I was officially one with the animals.

It wasn't just the porcine that surrounded me, either. The frequent neighs of Snickers the horse added considerably to the farm ambience. Deer and skunk were very common on the property, which increased my motivation to train Bugsy to have a reliable recall both for safety (deer) and sanity (skunks). I often saw rabbits, squirrels, and shrews, all of which provided tracking opportunities for Bugsy. As a canine professional, I recognized the excellent opportunities they provided for him to get both mental and physical exercise.

Almost as interesting to my dog as the animals living outside were the mice sharing our home. It's not unusual to have mice in the house out in the country, and that was especially true with this particular house. It was built in the mid-1840s, making it older than the state of Wisconsin, which entered the Union in 1848.

As such, it was no surprise that it was not well sealed and apparently had a sign written in mouse language that said, "Open to all. Welcome!" Bugsy did an excellent job of finding the mice. Usually, I was alerted to their presence by the sight of him play bowing to them in some corner or another. They surely felt trapped, but his mood was playful.

Old MacDonald would have loved this place, just as Bugsy and I did.

I miss living on a farm, though I am grateful that we have a lot of wildlife in our suburban lot in Flagstaff, too. We see skunks, foxes, deer, snakes, squirrels, and rabbits regularly and we even occasionally see a mouse in our house, apparently just for old time's sake.

Reindeer Are the Chosen Ones

December 23, 2014

I like to think of Santa Claus as a smart man as well as a generous one. Okay, so he's not known for brilliant nutritional choices, but since he is offered millions of cookies during a single day, he has challenges the rest of us can only imagine. On the other hand, when it comes to his entourage, he has chosen wisely.

He has a big job, but luckily, he has the magic of the season and the unflagging efforts of elves and reindeer to help. Elves are an obvious choice for Christmas work. They are highly skilled metalworkers, woodworkers, and artisans who work tirelessly and are the world's happiest labor force. But why did Santa pick reindeer? Why not stick with the Yule Goat of Norse tradition or the stork, which has long been trusted with the delivery of the most important gift of all—babies?

After some research, I came to understand Santa's wisdom. Reindeer (also called caribou) undergo an annual migration of thousands of miles, so one marathon journey in late December is within their endurance capabilities. These animals are able to run nearly 50 miles per hour, which is another way to say that they can really fly!

They won't scare the kids with weird calls because males only vocalize during fall mating season, and females only vocalize after their fawns are born in the summer. They're hardly silent, though. Some subspecies have knees that make a clicking noise while walking so they can stay together in a blizzard. That explains the need for Santa's sleigh bells. Otherwise that clicking would attract local reindeer and cause traffic jams.

Reindeer feet expand in summer when the tundra's ground is soft. They shrink in winter when the ground hardens, exposing the edge of the hoof which can cut into snow and ice to prevent slipping.

It's a good thing, too. Can you imagine the racket if all those reindeer slipped off your roof on Christmas Eve, dragging Santa and his sleigh with them? Those hooves allow them to dig through the snow to reach tasty lichen (also called reindeer moss) and in a pinch, they use them to uncover the occasional buried chimney, too.

During the Christmas season, a discussion of the color of reindeer body parts is far from unusual. Rudolph's red nose is truly the stuff of legend, but it has overshadowed another color issue because reindeer eyes change color seasonally.

The tapetum lucidum is the part of the eye behind the retina that reflects light to help animals see better when light levels are low. It's gold in most mammals, including reindeer in summertime. The seasonal color change to blue results in a greater scatter of light and therefore greater light sensitivity, though it causes a drop in visual acuity. This color change is unique among mammals.

Scientists consider this change an adaptation for dealing with the prolonged seasonal darkness so close to the North Pole, but have they really tested the possibility that it's just another part of Santa's magic?

Reindeer are truly the stuff of legend, but I love that their biology is as fascinating as the stories we tell about them.

So Many Species
January 6, 2015

My friends answered the ad for ducks, hoping to acquire some for their pond. When they went to check out the animals for sale, they were amazed to see Muscovy Ducks rather than the "real ducks" they were expecting. This type of misunderstanding often happens because there is far greater diversity than our language allows us to express easily. Confusion naturally arises about what the names of animals mean.

By "real ducks" my friends meant Mallards, which are the most common, but not the only, species of duck. In fact, there are about 120 species of ducks worldwide, so the term "duck" is only useful as a general way to describe an animal. The same goes for squirrel, which is often used as a definitive identification of an animal, though that's problematic because there are over 200 species of squirrels.

Even among animals that seem unique, there are often multiple species involved. For example, there are three species of elephants, five species of rhinoceroses, two hippopotamus species, and three zebra species. That's just a few of the biggest land mammals. In smaller animals, there are about 260 species of monkeys (plus about two dozen apes, including our own species), not to mention 90 deer species and 19 species of gazelles.

In the ocean, there are 42 species of whales, about 45 species of dolphins, 6 species of porpoises, 19 seal species, and 7 sea lion species. In salmon alone, there are 7 species, all of which are closely related to the roughly 20 species of trout swimming in the planet's waterways. (Just to confuse things, there are also 4 species whose common names include the term "salmon" even though they aren't really salmon.)

Even penguins and crocodiles are species rich, with 17 and 14 species, respectively. Crab diversity is at about 4,500 species, and that doesn't even count the 500 or so species of hermit crabs, which are not true crabs. That means there are more species of hermit crabs than octopuses, which account for around 300 species.

In the air, diversity is especially high because of the advantages flight confers. Bats are impressively diverse, with nearly 1,000 species. That is roughly a quarter of all mammal species. Lots of types of birds actually encompass multiple species. For example, there are upwards of 40 sparrow species, more than 60 species of eagles, around 270 hawk species, and an astounding 300-plus species of hummingbirds.

Though it's easy to be impressed by the 12,000 species of ants in the world or the 11,000 species of grasshoppers, beetles are the best ambassadors for diversity. It's silly to call this the age of mammals, when it is clearly—from an animal-centric perspective—the age of beetles, since there are nearly half a million different species of them. (From a life-in-general perspective, it is the age of bacteria, but that's another story altogether.)

My friends ended up adopting the Muscovy Ducks for their pond. I guess despite the diversity of the group, in some sense, a duck is a duck.

For many of the invertebrate groups, including insects, species numbers are greatly underestimated as many species remain undiscovered.

Egyptian Animal Mummies
January 20, 2015

Apparently, I'll never be past the stage of life where I write the "What I Did During My Vacation" essay.

A trip over winter break to the Washington D.C. area that included visiting the Capitol, the White House, and the Lincoln Memorial also allowed me the great pleasure of filling my brain to capacity at various Smithsonian Museums. As expected, my favorite was the Museum of Natural History, which I would rename the Smithsonian Museum of Heaven-For-Biologists if it were up to me.

I felt right at home surrounded by biology exhibits. Tom Eisner's prominently displayed quote, "Insects won't inherit the Earth—they own it now," reflects my own view, having completed a doctoral dissertation on insects. The dolphin and whale specimens suspended from the ceiling to show their true size felt familiar to me from growing up on the Southern California coast, diving and going on whale watching expeditions. The ideas in the "Once There Were Millions" exhibit about extinct birds such as the Passenger Pigeon, Great Auk, Carolina Parakeet, and Labrador Duck are everyday topics of discussion in my bird-crazed family.

In contrast, the animal mummies were definitely beyond my usual experiences with the natural world. Mummies are dead animals that have been preserved due to very specific conditions. (Skin and organs of deceased individuals only resist decay if exposed to specific conditions such as chemicals, extreme cold, low humidity, or hypoxia.) The term "mummies" usually refers to bodies embalmed with chemicals, but "accidental" mummies have been recovered from extreme habitats such as high in the Alps and in bogs. In Ancient Egypt, millions of mummies were made on purpose because a well-preserved body was considered critical for a good afterlife. Not all of them were human.

Egyptians mummified cats for multiple reasons: allowing a cherished pet to accompany a person in the afterlife, to provide food, or as offerings to the goddess Bastet, who was depicted as a lioness, a cat, or a woman with a feline head. Some cats were raised specifically for mummification. They were sold to people to leave in the temple as an offering after they worshipped a particular god. Many such kittens were 2 to 4 months old, and some experts believe this was simply because their small size made fitting them into mummy containers easier.

I knew that cats were held in high esteem by the Ancient Egyptians and even that they were buried with their owners, so the existence of huge numbers of cat mummies was not a big shock. On the other hand, it was surprising to learn that so many other species of animals were mummified. Common pet mummies included dogs, baboons, monkeys, gazelles, and mongooses. Crocodiles, fish, and bulls were also mummified in great numbers along with the occasional ram or hippo. According to some sources, the most commonly mummified animal was the ibis, with over four million discovered, making them many times more plentiful than cats, which are in a distant second place.

And that's the end of my vacation essay.

I have recently learned that the what-I-did-during-my-vacation essay is a relic from a bygone era. Kids are no longer asked to write such a piece, at least not in my area.

Identifying Individual Animals
February 3, 2015

My paint pen leaked enough green on the wasp in my forceps to cover her wings. She would never fly again, and that made me feel terrible. I was using a coding system of five colors and various areas of the wasps' bodies to number them up to 599. Occasionally, the equipment malfunctioned, which was frustrating to me and potentially lethal to my study subjects. Individual identification is essential in many behavioral studies, which is why I took the risk.

When studying migration, it is impossible to know how far animals go (and where) if you can't identify individuals. The same constraint applies to studies of home ranges and territories. Knowing who is who is essential if you want to know which individual is controlling a particular area, to understand the patterns of daily or seasonal movements, or to know whether individuals return to the same nesting site each spring. If you want to learn about an animal's lifespan, it is essential to monitor individuals from birth (or hatching) until death, and that can't be done without identifying them.

Individual identification is also necessary for studies of social groups. If you want to study the status hierarchy, you have to know which individuals are involved in every relevant social interaction. To know who is fighting and who wins each battle requires the identification of individuals. If you want to know who is mating with whom and whether they mate year after year or just once, you have to be able to recognize each individual. The same is true if you want to learn whether individuals specialize in their tasks (defense, foraging, caring for young, etc.) within a social group.

Animals can be physically labeled, and catching wild animals to place collars, tags, or otherwise mark them is common.

It's less invasive to use field identifications based on differences in appearance. If you look carefully, you can see that the splotches of color on the beaks of female Mallards vary from individual to individual. On killer whales (orcas), the unique patterns used to tell them apart are on the dorsal fins of both males and females. Humpback whales are identified based on the pigmentation and scarring on their tails and sometimes on their dorsal fins. For both orcas and humpbacks, photographic catalogs of individuals are assembled.

Technology has helped scientists to identify individual wild animals, especially with the wide availability and low cost of digital photography and videography. Techniques continue to improve, and new options are developed.

Facial recognition software is used on chimpanzees in the wild to know who is who based on both pictures and videos. Thanks to their stripes, specific zebras can be identified from photographs using technology remarkably similar to scanning barcodes. Many species have unique patterns, and by using 3-D surface models that are not affected by posture or camera angle, photographs can be used to identify tigers, giraffes, whale sharks, and cheetahs.

Such technology would have been beneficial to me, and even more so to my wasps.

Some methods of marking animals last long enough to be useful for that animals' entire lifetime such as tattoos or ear tagging. Other methods are temporary, such as fur clipping or dying.

Our Deadly Violent Relatives
February 17, 2015

Aggression is common in this species. They gang up on others and beat them up. They murder members of their own society and individuals from other groups. They wage war in organized assaults and commit rape. They assassinate their leaders. Does this sound like a species you know? I'm talking about chimpanzees, and the familiarity is no coincidence. They are our closest relatives, and share many traits with humans, including not always behaving peacefully.

The first evidence of the violent nature of the species came in the mid-1970s after researcher Jane Goodall's documentation of what became known as the "Four-Year War of Gombe" or the "Gombe Chimpanzee War." The conflict was between two groups of chimpanzees who occupied different areas of Gombe Stream National Park in Tanzania.

The southern group and the northern group who fought so hard for many years had originally made up a single social community. They split in 1974 into two separate groups, and were in a war until 1978. The southern group was made up of six females and their young, and seven males. The northern group was larger, made up of 12 females and their young, plus eight males.

In 1974, the serious violence began. Six males from the north attacked and killed a southern male while he was feeding in a tree. This was the first observation of chimpanzees deliberately killing another member of the species. During the remaining years of the war, the other six males of the southern troop were killed by chimps in the northern group. One southern female was killed, two went missing, and the remaining three were beaten, raped, and kidnapped by the northern group.

The long war between rival factions is far from the only example of chimpanzee murder in the wild, although it is one of the best documented.

Another recent example involves the death of Pimu, a leader who was killed by members of his own troop in 2011. It is rare for chimps to kill the leader, but it has been observed multiple times. Pimu had been leader for 7 years when he started a fight with the second-ranking male, who ran away. Other males in the group then attacked Pimu, biting him repeatedly until he died, which took a couple of hours. He was probably already dead when a former leader began smashing his head repeatedly with a big rock. Researchers described Pimu as an aggressive, unlikable, disagreeable, unpopular bully.

When Jane Goodall first reported on the war between these groups of chimpanzees, her observations were questioned. She was accused of both anthropomorphism and of causing any violence by her presence and behavior around the chimps, including providing them with bananas so they would tolerate her being there.

Recently, the idea that chimpanzee violence was an artifact of human presence and the resources they've provided has been discredited. A comparative study of many groups supports the idea that chimpanzees are naturally aggressive, and that they are, as some headlines have put it, "natural born killers."

One chimpanzee who was considered kind and peaceful lost his best friend (perhaps to violence from other chimpanzees but perhaps to human poachers) and he acted less social for many years. Then, later in life, he was a protector to orphan chimpanzees. I like to think that he lost his best friend to gang violence, rejected a lifestyle that served the purpose of territory expansion, and worked on behalf of troubled youth.

Symmetry, Beauty, and Mating Success

March 3, 2015

If you are particularly drawn to the looks of Brad Pitt, Halle Berry, Denzel Washington, or Nicole Kidman, then you'll be pleased to know that you are every bit as skilled at detecting beauty as a scorpionfly. The symmetry of many celebrities makes them attractive to other people just as the symmetry of scorpionflies makes them appealing to other members of their species.

Male scorpionflies compete with each other for access to females and then must be accepted as a mate by a female. Female scorpionflies find highly symmetrical males more attractive than those who are less symmetrical. Scientists studied the role of symmetry in male mating success, using wing length as the indicator of symmetry. The closer in length a male's forewings are to each other, the more successful he is at attracting females.

Does this make sense from an evolutionary point of view? In other words, are the females choosing higher quality mates when they select the most symmetrical males? The answer is yes.

Bodies of many animals have two sides that have been shaped by natural selection to be symmetrical. Deviation from that symmetry may indicate problems in development or exposure to environmental stress. Any departure from perfect symmetry may reflect poor health, lower survivorship, and lower reproductive capability.

Choosing symmetrical mates may be a way for females to select the healthiest partners. Health is often an indicator of high-quality genes, which are definitely worth choosing in a mate. Symmetry is linked to low levels of inbreeding and low parasite load, both of which are associated with health, vigor, and longevity. In scorpionflies, more symmetrical males as well as more symmetrical females have higher survivorship.

Insects such as scorpionflies are far from the only animals to judge mates based on symmetry. Studies have shown that attractiveness is linked to this trait in birds, fish, and mammals, including humans. People whose faces are more symmetrical are considered more attractive.

Symmetrical bodies are also rated as attractive, and men with symmetrical bodies tend to have typically masculine traits such as broad shoulders and smaller hip to waist ratios. Women who are highly symmetrical tend to have a more feminine, curvier shape. Human symmetry is correlated with better health and disease resistance, so choosing a masculine or feminine shape may be a way to choose a healthy partner with good genes.

Another way to pick a high-quality mate might be to take a cue from high school girls and date the quarterback. Men who play this position tend to be more symmetrical than men who play any other position. In the NFL, Matt Ryan had the highest degree of symmetry, though Brett Favre, Aaron Rodgers, Tom Brady, and Kurt Warner all cracked the top 10.

The Greeks didn't measure wing length in scorpionflies, and they certainly didn't worship the quarterback with the reverence they bestowed upon Zeus, but they did believe, long before science demonstrated it, that symmetry was a critical component of beauty.

Studies of scorpionflies were the first to demonstrate a preference by females for symmetrical males. Studies of these insects also found that females even prefer the pheromones produced by the most symmetrical males.

The Birds and the Bees . . . And the Pollen

March 17, 2015

If you've been suffering from allergies lately because of the juniper pollen, blame wind pollination and the copious amount of pollen it requires. Wind-pollinated plants must produce huge amounts of pollen in order for a few lucky grains to reach flowers of the right species and lead to fertilization.

Flowers that are pollinated by animals are far more likely to have their pollen end up on another flower of the right species because their animal pollinators deliver it directly rather than relying so much on chance. So, these species produce far less pollen than wind-pollinated species do. Animals visit the flowers and get pollen on themselves. When they visit the next flower, some of that pollen comes off.

Both the pollinators and the plants benefit. The plants receive pollination services, and the animals receive rewards, usually in the form of nectar. Pollinators and their flowers must be compatible for this exchange of goods and services to happen.

Bees are common pollinators. They are unable to see the color red, and the flowers they pollinate tend to be yellow, blue, and ultraviolet—the colors they see best—and have a mild sweet scent to attract them. Bee-pollinated flowers often have a landing platform that works for the bee species that pollinates it. For example, snapdragons open when a bee of the proper size lands on them, but bees that are too big or too small can't access the flower or the nectar rewards inside.

Hummingbirds see bright colors and are especially attracted to anything red. The flowers they pollinate are usually red with tubes that are the same size as their long beaks. Hummingbirds feed while hovering, and their flowers' petals curve away from the center of the flower. The shape of the flower causes pollen to collect on their head and neck while they feed. Hummingbird-pollinated flowers have no scent, which is not surprising as these birds have a poor sense of smell. Fuchsia, penstemon, and columbine are pollinated by hummingbirds.

Flowers pollinated by either bats or moths need to attract nocturnal animals with an excellent sense of smell. These flowers are white or pale yellow, which makes them more visible in darkness, and many only open at night. Moth-pollinated flowers such as morning glory and gardenia have a strong sweet smell and long tubes that match the length of the proboscis (similar to a tongue) of the moth. Bat-pollinated flowers smell musty and are large and strong enough not to be damaged by a hovering bat lapping up nectar and pollen. Bats pollinate mango, agave, and saguaro.

Butterflies have good vision but don't have a strong sense of smell. The flowers they pollinate are brightly colored but have little odor. They either provide a flat landing surface or occur in tight clusters that allow the butterflies to walk from flower to flower. Flowers pollinated by butterflies include milkweed, lavender, and aster.

The specificity of the matches between pollinators and flowers that makes pollination of flowers by animals possible is a biological, evolutionary marvel.

I appreciate the incredibly complex and fascinating nature of pollination by animals even more than usual when my eyes are itchy and watery from the excessive amounts of pollen being carried by the wind.

Elusive Giants
April 7, 2015

A giant squid's eye is larger than a human head, and the rest of its body is correspondingly immense. (An estimate of their maximum size is 600 pounds and 43 feet including 25-foot tentacles, so Pliny the Elder's estimate nearly 2,000 years ago of a creature with 30-foot arms that weighed 700 pounds wasn't far off.) Yet, this species is rarely seen and little understood. Though there have been reports of sightings at the surface of the water and even the occasional photo of a live one, most specimens have been found dead, washed ashore, or caught in commercial trawling nets.

The elusive nature of giant squids may account for tales of sea monsters, most notably the Kraken, attacking sailing ships. This legendary monster appeared in the mythology of many cultures. It was believed to live off the coasts of Norway and Greenland and to be large enough to swallow ships and all the sailors aboard.

Erik Pontoppidan, the Bishop of Bergen, described the Kraken in detail in the mid 1700s, and wrote that the real danger to sailors was not direct attack, but the whirlpools these animals created. He also claimed that fishing near them yielded bountiful catches because they were always near giant schools of fish. This is the origin of the expression, "You must have fished on Kraken" to refer to a good haul. Pontoppidan's description of the Kraken greatly influenced Jules Verne's account of the giant squid in *20,000 Leagues Under the Sea*.

It's remarkable that such a big animal was not photographed in its native habitat until the year 2004 and that the first video of one in the wild was shot in 2012. Though giant squids are not rare, they are solitary rather than being in large groups, and the ocean is vast. It's still a serious challenge for people to track them down because they are typically at depths of 2,000–3,000 feet.

An added impediment to recording them is that the equipment humans require to descend deep in the ocean is loud and bright, making it likely that squids will be scared away.

Much of what we know about the anatomy of this species is from specimens found inside the stomachs of their major predator, the sperm whale. Sperm whales are so good at finding their cephalopod prey that scientists follow them to find giant squids. These predators often have sucker marks and bite marks, presumably from giant squids attempting to fight them off. The beaks of giant squids are very hard, which is why they are the part most commonly found in the stomachs of whales.

Sperm whales are the only known predator of adult giant squids, although pilot whales may also eat them. (The juveniles have more natural enemies, and are often eaten by sharks and other fish.)

Giant squid behavior is largely unknown, reminding us that there are still great mysteries in our natural world. This is especially true in the deep ocean, which (for humans) remains the most inaccessible habitat on earth.

In 2019, researchers filmed a live giant squid in the Gulf of Mexico by dropping a lure on a line 6,000 feet into the ocean. The lure has faint blinking lights that mimic the bioluminescence of deep-sea jellies that giant squids eat. The giant squid approached the lure and touched it before swimming away.

Parasitoids Influence Host Behavior
April 21, 2015

The insects called "parasitoids" are poster children for the expression that truth is stranger than fiction. Parasitoids develop in or on the body of a host, often emerging in a way reminiscent of the famous chestburster scene from the movie *Alien*. They eventually cause their hosts to die, usually before they've had a chance to reproduce.

In addition to having catastrophic effects on the bodies of their hosts for their own benefit, parasitoids cause detrimental changes in host behavior. For example, one parasitoid affects mate choice in the crickets it attacks. In most species, including these crickets, females are more picky about who they mate with than males are, but parasitoids mess with that choosiness.

Female crickets who have been infected by a parasitoid fly are less choosy than females who have not been infested. Male field crickets sing to attract females, which usually prefer males who sing faster to males who sing at a lower rate. (Singing is an energetically costly and exhausting behavior, so males who can sing faster are advertising good health and fitness.)

Researchers found that females infested with parasitoids were equally likely to choose a slow singing male or a faster singing male, but uninfested females strongly preferred the faster singers. One theory is that infested females don't have long to live and are therefore choosing not to waste time and energy searching for the best mate at the risk of dying before they have a chance to reproduce at all.

Another parasitoid lays its eggs in a caterpillar, and even though the host is fattened up from eating, the dozens of larvae account for a third of the caterpillar's weight. They feed on the caterpillar from the inside out, but avoid eating any vital organs to keep their host alive. After some development, the larvae eat their way out of their host. It's as gross and repulsive as it sounds, because they use their saw-like mouthparts to cut their way out.

As they break through, the developing larvae release chemicals that paralyze the caterpillar, preventing it from doing anything to stop their exit. Ironically, a huge danger to the larvae is having another species of parasitoid lay eggs in their bodies at this stage. They spin silken cocoons around themselves, which offers some protection, and the parasitized caterpillar they have just used as a feeding ground and surrogate home also helps keep them safe.

If the caterpillar had not been a victim of these parasitoids, it would have spun its own cocoon in which to develop into a butterfly. Instead, the injured caterpillar uses its silken thread to spin a protective cocoon layer around the newly emerged pile of parasitoids. Its brain has been corrupted because of its parasitoid infestation, causing this change in behavior, along with another one: it defends the parasitoids against predators and other enemies by acting as a watchful parent until it starves to death.

Parasitoids are evolutionary sensations, but it's reasonable if you prefer to think of them as exemplars of the disgusting.

Parasitoids influenced the thinking of Charles Darwin, who famously wrote in a letter to the American naturalist, Asa Gray, "I cannot persuade myself that a beneficent & omnipotent God would have designedly created the Ichneumonidae [a family with many parasitic wasps] with the express intention of their feeding within the living bodies of caterpillars."

Mothers Are Givers
May 5, 2015

My husband was dissecting perch for a lab he was teaching. In the first fish he dissected, the heart, stomach, swim bladder, bladder, kidneys, liver, spleen, pancreas, and all other organs were laid out like a picture in a textbook, clearly visible and organized neatly within the body cavity. Sample one for the lab was complete! The second fish was a completely different story. The entire body cavity was packed with hundreds of eggs, and all the organs were shoved up toward the fish's gills.

I was 8 months pregnant at the time, and though my body had clearly swelled with child, the fish dissection provided the real aha moment for my husband: the baby growing inside me was taking up a lot of space and shoving my organs into an increasingly narrower space. I had already begun to picture my lungs being the shape of tortillas, and that's barely an exaggeration. The fish and I are fairly representative of mothers throughout the animal kingdom in that we contribute to the next generation with personal sacrifice and by giving of ourselves.

You would think that pygmy marmosets would have little to give, being the world's smallest monkey, but they, too, are generous with their body's resources. An adult female weighs about 4 ounces (113 grams). A female gives birth twice a year after a pregnancy that lasts over 4 months. So, she is pregnant more than 8 months each year, and every time she reproduces she has twins, whose combined weight of about 40 grams is more than a third of her weight.

Army ant queens have abdomens that are so distended for the purposes of productivity that they can lay up to 21,000 eggs a day. Termite queens can top that. They swell in size over a period of years until they are 500 times as big as a young queen, and can lay 40,000 eggs in a single day.

Elephant seal moms lose 600 pounds while nursing their babies in the first month after birth. They gain a lot of weight during pregnancy, but if you can lose the baby weight that fast and return to a pre-pregnancy size of 1,700 pounds, who cares?

Polar bears don't. A polar bear mom actually doubles her weight during pregnancy, and if she doesn't put on enough fat, her body will reabsorb the fetus. If she stays pregnant, she loses much of her fat stores during the 2 months that she spends hibernating. She's actually still hibernating during the birth.

Painful childbirth is common for mothers of many other species, but it's time to gain some perspective. Young sea lice eat their way out of their mother to be born. Any female who is not a sea louse should immediately experience feelings of gratitude.

Could that gratitude account for the fact that female orangutans never put their babies down during the first 4 months of their lives? Talk about being an attentive mother!

There's no doubt about it—mothers are givers.

Few people have the benefit of a fish dissection to make them even more empathetic to the experience of their pregnant partners, and I consider myself lucky as a result. The pygmy marmoset (whose name may be changed soon as it is offensive—see my note on page 150) was determined in 2018 to be made up of at least two distinct species and possibly more.

Camel Envy

May 19, 2015

Ahh, spring in Northern Arizona! This is the only place I've ever lived where it's the worst time of year. It's easy to tell that this season is upon us—skin that's dry and chapped to the point of bleeding, sunburns from walks to the mailbox, panic attacks if a water bottle is not within reach, and winds so strong you suspect you'll soon be on your way to Oz with Dorothy and Toto.

Another less universal sign of spring here is my annual envy of camels. I'm jealous that they are able to handle deserts far more harsh than what we face in our high mountain habitat. How do they do it? The camel's magic lies in many diverse adaptations that allow it to cope with tremendous levels of water stress.

Their physiology allows them to tolerate extreme dehydration. They can survive a level of water loss that would be fatal to other animals. Camels can lose up to 30 percent of their body's water, which is roughly a quarter of their body weight.

Camels can rehydrate quickly by drinking a large amount of water in one session. A thirsty camel can take in 130 liters of water in a few minutes and nearly 200 liters in a few hours. That fluid moves quickly to the bloodstream on the way to rehydrating the rest of the body. In most animals, such massive quantities of water entering the bloodstream so fast would be problematic because it would cause the rupture of red blood cells. The peculiar oval shape of camel blood cells as well as unusually strong, protein-rich cell membranes make them capable of swelling considerably and withstanding osmotic pressures that would be fatal in other species.

There are a number of adaptations that camels have to conserve water. They lose far less than other mammals through excretion of wastes. Their feces are exceptionally dry and their urine is very concentrated, containing little water but high quantities of salts and wastes. Camels even lose less water through respiration than many other animals do because they breathe slowly.

Though camels do sweat to lose heat through evaporative cooling, they minimize the need to do this in several ways. They are well protected from the sun and its dehydrating effects. The heavy fur and fat-filled humps on their backs protect camels from the solar radiation that would heat their bodies up. In contrast, the underside of the body has little fat or fur, so it can easily lose heat to the environment.

Camels can also tolerate huge ranges in body temperature. This conserves the water that would be lost through sweating and panting to cool the body to keep its temperature within narrower limits. Their body temperature changes throughout the day, starting out cool in the morning and rising as much as 6 degrees Celsius by the afternoon. Human body temperature fluctuates about half a degree.

I've heard of people wishing for the lovely eyelashes of the camel, but I simply covet their water conservation capabilities.

My camel envy grows greater each year, as do the droughts and water shortages in the west, which is not a coincidence.

You Are What You Eat
June 9, 2015

Animals must protect themselves from being eaten by other animals because (at the risk of stating the obvious) being eaten is rarely a successful evolutionary strategy. Being poisonous so that other animals can't safely eat you is a far better system. There are animals that produce toxins for this purpose, but some animals protect themselves from becoming someone else's dinner by ingesting poisons and sequestering them within the body. The poisons make them toxic to many potential predators.

Monarch butterflies are famously toxic, containing cardenolides—poisons that are toxic as well as bitter tasting, and cause vomiting in birds that try to eat them. These butterflies are among the animals that do not produce the toxins that protect them. Instead, they ingest these chemicals as larvae, which is when they feed on the milkweed plants that do produce them. They sequester those poisonous chemicals in their bodies, which makes them a poor choice of a meal for birds and other predators.

Poison dart frogs are among the most toxic of all animals. They owe their name to the practices of various

South American tribes of using the toxins to make poisonous darts with which they hunt. Like the monarch butterflies, these frogs acquire their toxins from their diet, particularly from mites, ants, and centipedes, who in turn acquire these toxins from plants that they eat. Poison dart frogs can ingest these insects, arachnids, and other arthropods without suffering ill effects from the neurotoxins, but very few animals share this ability. Similarly, few species can eat poison dart frogs without succumbing to the poison alkaloids they contain. One exception is the Rufous Motmot, which is capable of eating them without becoming sick or dying. This bird can neutralize the poison in its digestive system.

Another is the fire-bellied snake, which can detoxify the poison with chemicals in its saliva. In fact, this snake is so adept at handling toxins from poison dart frogs that it is the only known natural predator of the most toxic frog of all, the golden dart frog (*Phyllobates terribilis*). A single one of these frogs, which weighs less than an ounce, produces enough toxin to kill 10 to 20 people. It takes just 136 micrograms of the toxin—roughly the amount of 2 or 3 grains of table salt—to kill a person of average size.

Captive-bred poison dart frogs are harmless. Wild-caught ones that are kept in captivity eventually lose their toxicity. This happens because a diet of crickets and fruit flies does not provide them with the toxins that are necessary for them to maintain their poisonous status. (These frogs typically won't eat ants in captivity, perhaps because the species of ants they eat in the wild are not commercially available.) Moms of some species of poison dart frogs lay unfertilized eggs that contain tiny amounts of these toxins for their tadpoles to eat, suggesting that even the youngest members of such species are chemically protected.

For many toxic animals, you truly are what you eat!

Sequestering toxins from food is most common in invertebrates, especially in arthropods and mollusks. It's rare in vertebrates, but in addition to frogs, there are a few snakes and birds capable of doing this.

Fatherhood Lessons from the Red Fox

June 23, 2015

I once commented to a high school friend that parenting is so NOT what it looks like in the brochure. This mother of five replied, "You got a brochure? You're so lucky—I didn't." Most lessons on parenting are learned as we go. That's a shame because there's a lot of wisdom worth passing on, sometimes from unexpected sources.

Consider the red fox. In any list of exemplary dads, this species is likely to make an appearance, and with good reason. He follows some fundamentally sound principles of fatherhood.

Take care of the mom so she can take care of the babies. In the first weeks after a litter of fox kits arrive, the mother stays with them constantly. They need her body heat to maintain their own temperatures in the correct range. During that time, the father takes care of both the mom (the vixen) and the kits, bringing them food several times daily. They are not just getting breakfast in bed, but other meals and snacks, too.

Spend time with your children. Red fox dads do more than just deliver food, although that alone is an important contribution to the survival of the young. These dads are actively engaged with their kits as they develop. In the early months, they lead the young around their territory, helping them learn their way around and alerting them to potential dangers.

Change your behavior as your kids grow. When the kits are about 3 months old, red fox fathers stop bringing food to their young, but they still contribute to their sustenance. They hide food on their territory, which provides the kits lessons in using their sense of smell to track down food. They also teach the kits how to hunt, starting with the prey that even young foxes can hunt successfully such as rabbits, game birds, and rodents.

Play together. Red foxes are attentive dads, often playing with their children. Play is an important part of parenting because of the many benefits to the young. Through play, young foxes practice adult behavior such as hunting, fighting, and courtship rituals. They also learn how to recover when their bodies are off balance or out of control and develop the emotional skills necessary for handling unexpected stress. Oh, and they have fun, too.

Be prepared to assume all parenting duties. If the vixen dies, the father takes over the care of their kits. It's easier to raise young foxes with two parents, but like single dads everywhere, he is perfectly capable of doing a good job. The younger his kits are when the mother dies, the greater the challenge.

Refuse no offer of help. Whether a single parent or a pair is caring for their young, foxes usually have help. Red foxes live in family groups, and other members of the group, including kits from previous years, help care for the young.

Red foxes make great dads, even without the benefit of any brochure.

I hope everyone had a happy Father's Day!

I don't often give other parents advice because it's irritating, but I do like to share with new parents the importance of accepting offers to help, like foxes do. It's one area where parents in our culture sometimes need a little push for the benefit of themselves and their children.

Ocean Love
July 7, 2015

The sea lion put his head on the board of my windsurfer, and it's fair to say that I have rarely been so scared. It's hard to overstate the startle factor of something suddenly and unexpectedly surfacing from the ocean. California sea lions are huge animals with big jaws and teeth, and I felt pretty vulnerable a quarter mile off the shore of Catalina Island. I was drifting offshore waiting for the wind to pick up enough for me to head back to the cove I called home at the time.

After the initial shock, I enjoyed the brief time that the sea lion was so close to me, relishing the opportunity to see a member of one of my favorite species and share a nature moment few people have been lucky enough to experience.

Years ago, I told this story to my friend Laura—a fellow dog trainer—at a conference about applied animal behavior where we had just met. Laura began her work as a professional animal trainer working in aquaria with marine mammals. We bonded over our love of the ocean and shared scuba diving stories ranging from seeing sharks to swimming with seals. The conversation turned to our wish lists of marine experiences.

Karen: I'd love to see a sea turtle coming ashore to lay eggs.

Laura: I've seen that! It was really amazing.

Karen: Ooooh, you're so lucky! I've seen the very recent tracks of a turtle, but all that told me was that I had just missed it. (Several years after this conversation, I saw several sea turtles laying eggs on the beach, and it was as incredible as I had imagined.) And I know you've experienced my biggest wish—to swim with dolphins. I've been in the ocean with sea lions, seals, and even sea otters, but not dolphins. I've seen them alongside big and small boats I've been in, but have yet to swim with them.

Laura: What I'd most love to see is a blue whale.

Karen: How cool would that be? I once saw the spout of a blue whale. Obviously, that's not as cool as seeing the whale itself, but I consider myself extremely fortunate just to have seen the spout.

Laura: A friend who I met during a wild dolphin research project is even more blessed. One day in her office, I saw the most magnificent photo of a blue whale swimming a few feet underwater alongside their research boat. In the photo, she is standing on deck looking at the whale. They were doing blue whale surveys and the aerial photographer who was documenting each sighting gave her the photo as a gift.

Ocean lovers that we are, we both agreed it's truly the photo of a lifetime—the one thing you'd grab out of a burning house once all the living creatures were safely accounted for. And we both paused to think about our blue whale sighting envy and how much we love all the creatures in the ocean.

The Laura in this story is well-known dog trainer Laura Monaco Torelli and she is most definitely the first person I contact to share any marine adventure I am lucky enough to have.

Animals Use Tools, Too

August 4, 2015

Fifty years ago, tool use was considered a distinctly human trait, defining our species as "Man the Tool Maker." Many people believed that our ability to make and use tools distinguished us from animals. It was common at that time not to recognize that humans are animals, but what else could we be? Plants? Fungi? Bacteria?

Then, in the early 1960s, scientist Jane Goodall observed chimpanzees using tools. She saw one individual "fishing" for termites by poking grass into termite mounds, and then eating the termites that were on the grass. She also saw chimpanzees pulling leaves off twigs to create a stem better suited to such fishing for insects. The discovery of another species doing what was thought to be possible for humans only led to this famous remark to Goodall in a telegram from anthropologist Louis Leaky: "Now we must redefine tool, redefine Man, or accept chimpanzees as human."

The idea that tool use was unique to humans was a case of, as T.H. Huxley said, "The great tragedy of Science—the slaying of a beautiful hypothesis by an ugly fact." Soon, there were more ugly facts to deal with because many species were observed using tools. They weren't all primates, either.

Sticks, branches, and leaves are the most common tools in the animal world. Elephants use branches as back scratchers and leaves as fly swatters. Gorillas measure water depth with walking poles before wading in. The Woodpecker Finch of the Galápagos Islands uses twigs to "fish" for insects, much as chimpanzees do. Ants use leaves to carry food and water to their nests.

Rocks are also used by a wide variety of species. Egyptian Vultures use them to crack open ostrich eggs, while various wasps use pebbles to smooth out mud or sand during nest construction. Sea otters pound urchins with rocks to crack them open, and capuchin monkeys do the same to break nuts open.

Beyond these relatively basic uses of tools by animals are more creative and complex examples of tool use. Dolphins use sponges to protect their noses from abrasions when they are foraging on the bottom of the ocean. Blanket octopuses are immune to the venom of the Portuguese man o' war, but they will rip off the tentacles of these jellyfish and use them to defend themselves against attackers.

Cockatoos use leaves to help them open nuts just as we use rubber jar openers to open containers of food. Such tools make it easier for these birds to twist open nuts and for us to twist open peanut butter or pickle jars.

The fact that tool use occurs in a wide variety of animals doesn't make it any less amazing. I don't want to hear people disparage tool use because "even animals" do it. I'd like to hear people be so impressed by any examples of tool use that they exclaim, "Wow! Tool use in humans!" with the same degree of enthusiasm usually reserved for observing the prowess that other animals exhibit with tools.

In addition to their use of tools, humans have been considered unique because they have language, create art, cooperate with others, are capable of deception, and for many other reasons. In each case, animal examples have been discovered that wreck the idea that humans are unique. As Charles Darwin wrote, humans and other animals differ only in degree, not in kind.

Animals Can't Do What?
September 8, 2015

Everyone has curious gaps in their abilities. In my case, for example, I can't take in information about epic movies. It was 20 years after the release of *Star Wars* that I realized that Darth Vader was Luke Skywalker's father. (Frankly, it's a wonder my husband still married me.)

Many entire species of animals are unable to do things that it would seem anyone could do. Elephants can't jump at all as adults—not even an inch. The bones in their feet are so close together that their feet lack the flexibility or spring mechanism to generate enough force. Because they weigh so much, the force required for a jump would be so high anyway that it's hardly surprising that elephants are incapable of doing it. They are not the only animals who can't catch air. Hippopotamuses and rhinoceroses can't either.

Rats can't vomit. The barrier between their stomach and the esophagus is extremely strong. It's so powerful that the surrounding muscles do not have enough strength to open it by force, which is required for vomiting. Additionally, their diaphragm muscles are unable to contract independently, which is also necessary for vomiting. For similar reasons, they lack the ability to burp and generally do not experience any acid reflux.

Most fish can't close their eyes for a very simple reason: they don't have eyelids. In most animals, eyelids play an important role in keeping eyes from drying out. Since fish live in water, they don't need eyelids for this purpose. A few fish, including jacks, mackerel, and mullet, have immovable transparent adipose eyelids that can cover part or all of the eye. Nobody knows for sure what they do, but scientists have several theories. They may act like a lens, provide a protective barrier, block ultraviolet light, or allow fish to see polarized light. Most sharks have clear eyelids that can be moved horizontally across their eyes to protect these delicate organs when they are feeding on prey.

Crocodiles can't stick out their tongues. A membrane keeps the tongue attached to the bottom of the mouth. I guess they have to insult other members of the species by fighting them to the death or something.

Horses can't breathe through their mouths. They can only take air in through their noses. Most mammals raise their soft palate to allow air from the mouth to travel to the lungs through an open epiglottis. Horses have an especially long soft palate, so in order to move it out of the way, the epiglottis has to move down, which makes it cover the windpipe. So, when the path from the mouth is open, the path to the lungs is closed.

I can do all these things that these other animals can't do, so I'm tempted to be smug about it. However, in addition to my *Star Wars* lapse, I have a truly dreadful sense of direction. So while I know I shouldn't point fingers at other animals, please note that I am perfectly capable of doing so.

Abilities that are essential for some species are irrelevant for others, making the fact that there are differences in what each one can and can't do predictable in addition to interesting.

What's Up with Giraffe Necks?
September 22, 2015

The common response to the question, "Why are giraffe necks so long?" is that giraffes evolved long necks because they allowed them to feed on higher leaves in the trees. Most people accept it without much thought, but that doesn't make it true.

It's certainly true that giraffes are able to reach higher leaves than shorter animals, but that doesn't mean that this capability was a driving force in evolution. Giraffes actually spend very little time foraging on high leaves in trees, feeding primarily at shoulder or even belly height. When droughts make food scarce and competition for it is most intense, they eat low brush. During their evolutionary history, giraffes have grown proportionally more in their necks than in their legs, which is a physiologically costly way to increase height. These observations suggest that the advantage of reaching higher leaves is insufficient to explain the evolution of the giraffe's long neck.

Just because an animal uses a feature in a particular way is no proof that it evolved for that reason. It's common in evolution for a trait to be co-opted for a different purpose. Feathers in birds initially evolved for heat regulation, but later served two very different purposes: displays and flight. The lungs of ancient fish evolved into something quite different in the lineages of modern fish. In these derived fish, the air sacs have become swim bladders, which allow fish to control their buoyancy.

Evolutionary biology, like any field of study, has its share of jargon, so as expected there's a specific term for a trait that has been co-opted for a different purpose than the one it performed originally and for which natural selection shaped it. It's called an "exaptation." The term was considered necessary because a current function of a trait does not necessarily explain how and why it developed. An "exaptation" is different than an "adaptation," which is a trait whose current function evolved by means of natural selection and is maintained by it.

The long necks of giraffes are likely an adaptation that increases a male's chances of defeating other males in fights over females. Male giraffes use their heads and necks as clubs against each other, causing injuries and occasionally killing each other. Their necks are larger and more heavily armored than females, who don't fight. Males with longer necks win fights more often and attract more females as mates.

Males have longer necks than females, and suffer more predation than females do. Because they are taller than females, the challenges of managing their blood pressure and of drinking water by splaying their front legs or kneeling in order for the head to reach the ground are also greater. The giraffe neck is comparable to the peacock's tail in that it promotes more mating success while also exposing the males to dangers and physical challenges.

Multiple types of evidence support the conclusion that the giraffe's long neck is an adaptation for fighting between males and an exaptation for feeding on leaves high in the trees.

The long neck of the giraffe is also likely an exaptation for being a lookout, making it easier to keep watch over a long distance for potential predators.

America's Speed Demon
October 6, 2015

"Where the deer and the pronghorn play" may sound funny to most of us, but it is more accurate than the original lyric.

Pronghorn are known by many names—prong buck, cabri, pronghorn antelope, and antelope—but they are NOT antelope. They are only called by that name because they resemble the antelope of the old world and fill a similar ecological role.

Pronghorn are more closely related to giraffes and okapi than to antelope. The resemblance of pronghorn to antelope is not due to common ancestry. Rather, the matching traits are a result of convergent evolution, which is the process of unrelated organisms evolving similar traits because of pressures to adapt to similar environmental conditions.

A well-known example of convergent evolution is the wings of bats and birds, which evolved separately in different lineages. Another such example is the eye of the octopus and the eye of vertebrates, which also evolved independently, but look quite alike. In both of these cases, modern species share a trait that their common ancestor did not.

Both antelope and pronghorn are fast because they evolved with pressure from fast predators. Pronghorn can run nearly 60 miles per hour for short bursts, and they can maintain speeds of around 30 miles per hour for many miles—longer than any other runner on the planet. Their running ability allows them to outsprint potential predators in the same way that antelope use their speed to avoid becoming cheetah chow. Only cheetahs are faster than pronghorn, which are the fastest land mammal in North America.

That speed is a result of the need for pronghorn to run away to escape predators. It is thought that this speed evolved in response to the running ability of predators such as the American cheetah that are now extinct. (Interestingly, the American cheetah is actually more closely related to mountain lions than to cheetahs, but evolved many of the same characteristics of its namesake through convergent evolution.) The speed of pronghorn is beyond what's necessary to outrun the predators such as bobcats, wolves, and coyotes that are currently in its range.

That's why scientists have postulated that the faster extinct predators played an important role in turning pronghorn into the Usain Bolts of America's Great Plains.

In addition to their impressive speed, pronghorn have exceptional eyesight. That's another adaptation to life in an open habitat that offers no place to hide. Spotting predators from a distance is obviously advantageous, so there was tremendous evolutionary pressure for seeing well and seeing far. Humans need eight-power binoculars to see as well as pronghorn, which is why they so often see us before we see them.

The scientific name, *Antilocapra americana*, means American goat-antelope. While the "goat-antelope" part is a bit off the mark, the "American" in its name is spot on. Their range extends from southern Canada to northern Mexico, and they are primarily found in the dry open areas of the western United States, meaning that pronghorn are a treasure found only in North America.

Pronghorn have 13 distinct gaits, and one of those involves strides of over 7 meters. In contrast, horses have 5 natural gaits and 4 additional ones that can be taught with varying degrees of success depending on the breed. Seeing pronghorn regularly is a huge perk of living in the southwest.

Sponges Are Animals
October 20, 2015

While few people have trouble correctly classifying horses, frogs, sea stars, and worms as animals, it's harder to recognize that certain other life forms belong to this same kingdom.

Among the organisms that are not always properly identified as animals are sea anemones, which look more like flowers to most of us. Another example is the sea jellies, which resemble floating umbrellas rather than stereotypical animals. (These are the animals formerly known as jellyfish, but the name sea jellies is considered more appropriate since they are not actually fish. The fact that they are not actually jelly seems not to bother those in charge of the world's scientific nomenclature, but it certainly bothers me.) Coral can appear more like a mineral than an animal or a vegetable, but they, too, are animals.

Perhaps most counterintuitive are the sponges, which resemble, well, sponges, more than anything else. So, why are they considered animals?

Simply put, they are classified this way because they possess the defining characteristics of animals. As I'll explain, they are multicellular heterotrophs made up of eukaryotic cells that lack cell walls.

Being multicellular just means that they are made up of many cells as opposed to being single-celled organisms. This is a trait that animals share with plants, fungi, and many of the protists.

A heterotroph is an animal that must get its energy from other organisms. Their food is the fats, carbohydrates, and proteins from other living things. They can't make their own energy-containing organic molecules from inorganic raw materials and an energy source such as sunlight, as plants do via the process of photosynthesis.

Eukaryotic cells contain membrane-bound organelles including the nucleus, while prokaryotic cells do not. Also, they can be hundreds of times bigger than prokaryotic cells. Bacteria and archaea are single-celled organisms and those single cells are prokaryotic. All multicellular organisms and the single-celled protists are composed of eukaryotic cells.

Animals' cells lack cell walls. Fungi have cell walls made of chitin and plants have cell walls made of cellulose. Some protists have cell walls and some don't.

That's it! Any organism that is a multicellular heterotroph made up of eukaryotic cells that lack cell walls is an animal. They do not have to have a mouth, be able to move, be symmetrical or contain tissues and organs, though many animals do.

So even though sponges look superficially like plants or even like blobs, they are actually animals. Many people have trouble believing this, mistakenly thinking that they are plants. The main trait that they share with plants is that they are sessile, which means that they are immobile and fixed to one spot. Other animals such as barnacles, tunicates, and some marine worms are also sessile—stuck to a substrate such as a rock or a part of a coral reef.

Aristotle found sponges confusing, considering them somewhere between a plant and an animal. In the 1800s, after centuries of debate during which they were usually considered plants, it was finally understood that sponges are animals.

Though many people understand the general category of animals, it continues to amaze me how many people think that only mammals are animals. I regularly hear people ask if ants are animals or if birds are animals. Yes, yes they are!

Biomimicry: Nature's Solutions

November 3, 2015

The process of natural selection has resulted in effective biological solutions to dilemmas faced by people. Taking inspiration from these proven successes and applying it to our own problems is called biomimicry. It's an innovation strategy that seeks answers to human challenges by copying nature's solutions to similar problems.

One problem solved with biomimicry is the violation of Japanese sound ordinances by high-speed bullet trains. Maintaining speeds of 200 miles per hour requires that trains travel on straightaways rather than on curves, and that necessitates going through obstacles and not around them. The air compression caused by trains in tunnels was causing tunnel booms as they emerged. (Tunnel booms sound like sonic booms, although the physics behind these two phenomena are different.)

Engineers were tasked with lowering the sound without compromising speed. The chief engineer was a birder who observed that kingfishers dive at high speeds from air into water with almost no splash. That observation led him to a solution because the same form allows trains to pass through tunnels while minimizing air friction. That's why the fronts of bullet trains are shaped like the heads of kingfishers.

Bats use echolocation to navigate, and now blind people can, too. The design of the "bat cane" is based on echolocation—sending out ultrasonic pulses and listening to the pattern of the sound waves that bounce back. The cane releases 60,000 ultrasonic pulses per second. When sound waves bounce back especially quickly, it's an indication that there's an object nearby, and the cane's handle vibrates to let the person know.

The clear wings of glasswing butterflies reflect almost no light. Wouldn't that be a great feature in our handheld devices? Work is underway to develop surfaces for phones and tablets that are similarly non-reflective. By examining the clear portions of butterfly wings under an electron microscope, scientists learned how butterflies minimize reflection in order to make themselves less visible to potential predators.

The secret is a series of tiny pillar-like structures of varying heights that are spaced randomly (rather than regularly) on the surface of the wing. This erratic arrangement results in 2 to 5 percent of light reflecting, compared with 8 to 100 percent for flat panes of glass.

Porcupine quills go in easy, but are tricky to remove, as so many dogs have discovered. The barbs on the quills grab at tissue as they are pulled, causing pain and damage. This flesh-grabbing quality inspired a way to achieve a common medical goal—closing wounds, surgical incisions, and hernias. Medical tape with microscopic barbs may seal these injuries more effectively than staples and various mesh products currently in use.

Humans engineer many incredible technologies, but only the models, prototypes, and innovative designs based on the way nature did it first have the direct benefit of eons of evolution. As Michael Pawlyn, a leader in biomimicry, says, "You could look at nature as being like a catalog of products, and all of those have benefited from a 3.8 billion year research and development period."

Biomimicry examples abound, whether that means using efficient whale fin shapes as models for wind turbines or designing surfing wetsuits that trap warm air like beaver fur does. In so many cases, animals inspire great engineering design.

Blue Whales Feast Every Day
November 17, 2015

If you think you and your family are champion eaters based on your Thanksgiving feast, you may not fully realize how poorly you stack up next to the competition. You can boast that you eat "a ton" on Thanksgiving, but blue whales have you beat.

A single individual can eat four tons of food—no quotation marks—because this is literally what they eat in a day. That means they're taking in some 8,000 pounds of food, making your Thanksgiving intake look like you weren't even trying.

Blue whales can dive to depths of 500 meters below the surface to feed, though typical dives go to 100 meters. They commonly stay underwater for 10 to 20 minutes. These animals dive repeatedly during the day, but are far more likely to feed on the surface at night.

They are the largest animals ever to live on Earth, with lengths approaching 30 meters and weights of 170 tons. Even their tongues are massive beyond imagination. At 2.7 tons, a blue whale's tongue is bigger than a rhinoceros or a hippo, and similar in size to a small adult elephant.

Blue whales need enormous tongues because of the unusual way that they eat. These giants are not eating other large creatures of our seas. On the contrary, their diet consists of small organisms, mostly krill. Krill are crustaceans, superficially similar to shrimp, that live throughout the world's oceans and are usually about 1-2 cm long. A blue whale may eat up to 40 million krill in a day, and the giant tongue helps them do this.

A blue whale swims at high speed through masses of these crustaceans with its mouth wide open. The force of the water pushes its mouth back, increasing the space available to hold water. A blue whale can take in 90 tons of water in a single gulp. Once its mouth is full, the whale squeezes the water out with its tongue, while trapping the krill and other organisms from that mouthful of water with a structure called baleen.

Baleen consists of plates of keratin with fringes that hang down from the whale's upper jaw into the mouth. It acts like a sieve, allowing the water to escape, but preventing krill and other small organisms from passing through. After the water has been forced out, the blue whale swallows the food that is left behind. The baleen whale group takes its name from this feeding structure, and includes blue whales, right whales, minke whales, gray whales, and humpback whales.

It's easy to see that humans aren't that impressive when it comes to eating. Blue whales, on the other hand, really know how to overeat, and they don't just do it on holidays and special occasions. They are the everyday feasters who show the rest of us how it's done. So, go ahead, add some humble pie to those slices of pumpkin and apple pie, because on the blue whale scale, what you eat on Thanksgiving Day doesn't even register.

At one Thanksgiving dinner when I was in graduate school, each of the seven of us brought the type of pie our family considered most essential. While I was uncomfortably full following my poor decision to eat a piece (albeit small) from each pie, I accepted how poorly I stacked up to the consumption habits of blue whales.

Insects of the Snow
December 8, 2015

Are you crying your eyes out because the wintry weather means no insects until spring? Fear not and get ready to celebrate if it's news to you that there are insects that live in the snow, including snow scorpionflies, snow fleas, snow flies, and ice crawlers.

Don't get all crazy, though, by bringing them inside to adopt as pets. They are so well adapted for life out in the cold that they can't handle higher temperatures. Some of them, such as snow scorpionflies, can even be killed by the exposure to warmth if held in a human hand. These insects are most active in late winter, and feed on mosses. They travel between breeding areas by walking across patches of snow.

Snow fleas aren't really fleas, and according to recent taxonomy work, they aren't even insects, although they used to be classified in that group. They are springtails, which are primitive hexapods closely related to insects. Their name comes from their ability to jump by tucking their own tail-like appendage beneath the body under high tension. When released, it pushes against the substrate, launching the animal into the air.

The protein in the bodies of snow fleas that allows them to function in temperatures below freezing is glycine-rich and unlike any other known proteins. It is possible that researchers will soon accomplish their goal of developing similar proteins that can prolong the viability of organs to be transplanted. As this protein prevents the formation of ice crystals, it may provide a way to store organs at lower temperatures, which makes them last longer. Though possibly of less importance to some, this protein may also be useful in making creamier ice cream.

Snow flies are a type of crane fly that can be seen walking on top of snow. They have glycerol in the fluids in their body, and that protects them against freezing. They drink water directly from snow, and like many flies, they feed on feces—in this case from rodents. In fact, they spend a lot of time in the tunnels of small mammals, which can protect them from the most extreme weather. Other refuges of snow flies include under leaf litter, deep inside caves, and in pockets of air made by snow landing on top of grass and other vegetation that is bent over.

Ice crawlers make up the very unusual family of insects called Grylloblattidae whose members are mainly found at high altitude, high latitude, or on glaciers. They, too, are adapted for life on snow and ice and in the cold, and are the best known of the insect psychrophiles—the technical term for an organism that is capable of growth and reproduction in cold temperatures (minus 20 C to 10 C). For anyone interested in learning more jargon, psychrophiles are a specific type of extremophile, which is an organism that thrives in extreme conditions that are detrimental to most life forms.

Isn't it great to know that some insects can get in on the winter fun?

Before I moved to Flagstaff from New Hampshire (and Wisconsin before that), I enjoyed having a cold season that seemed virtually free of insects, as I did not miss the biting ones like black flies, deer flies, and mosquitoes so common in the summer. Now that I live at high altitude where there are so few biting insects, I find that during winter I do miss the insects of the other seasons.

Tern the Table on Santa's Reindeer

December 22, 2015

Santa Claus' reindeer are famous for flying all over the world, and because they do it in a single night, they get all the glory. That's not fair to the Arctic Tern, a species that travels across the planet without the benefit of any magic.

The Arctic Tern has the longest annual migration of any animal. Members of this species fly from Greenland to Antarctica and back each year. Their travels have been harder in recent years because climate change has forced many birds to travel further north to breed, resulting in more deaths en route.

Scientists recently learned that these birds don't take the shortest route straight up the center of the Atlantic, as logic might suggest, but travel by zigzagging. On their way north, they leave Antarctica for Africa, then fly to South America, and finally head to the Arctic. The indirect S-shaped path could be a strategy to avoid flying into the wind by working with global wind systems, which move clockwise in the North Atlantic and counterclockwise in the South Atlantic.

On the trip south in the fall, Arctic Terns have a lengthy refueling stop in the Northern Atlantic. Birds spend nearly a month feasting on fish and zooplankton before heading south across the tropics. In total, the southern migration takes 3 months, while the journey north takes less than half that time. Such details of the migration are known because scientists were able to fit these 110-gram (4-ounce) birds with transmitters weighing less than 1.5 grams (about a twentieth of an ounce) to track them. Such tiny technology has only been available recently. This research has shown that the long migration—already known to be nearly pole to pole—is thousands of kilometers longer than previously thought.

Individuals travel 71,000 kilometers (44,000 miles) a year, and though they travel light compared to Santa's reindeer—no sleigh, bells, gifts, or portly passenger—the energy requirements for their journey are enormous. That provides a likely explanation for why their route takes them through highly productive areas of the oceans. The migration routes of Arctic Terns tend to involve travel through areas of the ocean that are high in productivity, which means that they are rich in food sources. Polar regions of the ocean offer greater quantities of food, which may explain why migrating Arctic Terns spend so little time in the temperate and tropical regions of the Atlantic.

Unlike Santa and his reindeer, who spend all year at the North Pole except when they are flying about on Christmas Eve, Arctic Terns live in perpetual summer. They spend the breeding season in the Arctic and sub-Arctic parts of Asia, North America, and Europe, and the northern hemisphere's winter in the Weddell Sea on the shores of Antarctica. Maybe the fact that they experience more daylight each year than any other species has made people too jealous to celebrate their journey as we do for reindeer. Or perhaps delivering gifts and spreading joy accounts for the excellent press enjoyed by Rudolph and his friends.

Santa Claus liked this article, and according to my sources, he has reached out to the Arctic Tern community for some travel and fueling tips.

Acknowledgments

A book of columns involves the work of many people over many years, and I have immense gratitude for all the people who played a role in developing my interest in animals and in writing.

Thank you to my dissertation advisor, Bob Jeanne, who taught me so much about the fundamentals of animal behavior and how to conduct research in this field, and whose great love for social wasps is contagious. I'm also grateful to the other members of my dissertation committee for their great enthusiasm about animals in general and their passions specifically: Jeff Baylis (fishes), Jack Hailman (birds), Dave Hogg (insects associated with field crops), and Dan Young (beetles). I learned so much from all of them.

Thank you to the *Arizona Daily Sun* for allowing me the freedom since 2008 to write about all kinds of animals in what was originally the newspaper's pet column. I'm especially grateful to the managing editors of the paper—the late Randy Wilson and Chris Etling—for their ability to keep the newspaper going strong through such tough times. I appreciate the many editors who managed my column over the years, including Arlene Hittle, Svea Conrad, Larry Hendricks, and Sam McManis. Thanks also to Sheila Madrak, who wrote the column before me and encouraged me to take it over when she moved. May any column that mentions turtles be a tribute to her love for all species in the order Testudines!

I am lucky to have been raised by such wonderful parents, and fortunate they doubled up as editors for this column. My Dad, Ralph London, is a man of great intellect. He is a scholar with a curious mind, a brilliant sense of humor, a positive enthusiasm for life, and the willingness to edit each column before I submitted it and offer helpful suggestions, large and small. He is the source of many of the best qualities in both his daughters.

My Mom, the late Bobbi London, was unbelievably dedicated to motherhood and prouder of her daughters than it is possible to describe. No compliment brings me more joy than when someone says I remind them of her. I do my best to follow her example of being strong-minded, clever, giving, caring, helpful, industrious, and attentive. She never let an errant comma—or any other literary or grammatical irregularity—escape her notice when she edited my columns, and was wonderfully appreciative of every animal I ever brought home, including the snakes, frogs, ants, and stink bugs that she quite rightly thought should stay outside.

I'm grateful to be the younger sister of Marla London, who has always been a good writer, and who inspired me since early childhood. There's something extra special about having another person who grew up in the same household as me who is the only one truly capable of understanding my life and world in some ways. She's literally the only other person on the planet who knows the names of all the stuffed animals we had as kids, and that's pretty special. (I'm looking at you Goggy, Old Ratty, New Ratty, Tigger, Puff Puff, and Codenly!) On top of that, she so kindly agreed to proofread this entire manuscript, thus saving me from the worry of too many mistakes appearing in the final version. Though I am usually happy to share with my sister, I'm taking a "mine, mine, mine" approach to all the remaining mistakes, as they belong entirely to me.

Thank you to my husband, Richard Hofstetter, and our sons Brian Hofstetter and Evan Hofstetter, who have all offered many ideas for columns, read over my columns and offered comments prior to submission, and assured me that my jokes were amusing, whether or not that was, strictly speaking, true. So many of the adventures I write about involved these three members of my family, and those memories are the most special treasures. Their permission to write about these experiences and share details about their lives so publicly is an act of kindness I receive with extreme gratitude.

I am so grateful that Melissa Hafting allowed me to use her wonderful photograph of an Arctic Tern on my cover. The design would not have felt right without it!

Thank you to my editor, Eileen Anderson, who shares so generously her knowledge and expertise about grammar, writing, animals, and publishing. I know how much better this book is because of her input, and for that I am profoundly grateful. Just as importantly, she is a good friend who brings joy and laughter to every interaction. She is always down for a Zoom call when I need an "Eileen fix," whether for personal or professional reasons.

About the Author

Karen B. London, PhD, is a Certified Professional Dog Trainer and a Certified Applied Animal Behaviorist. She received her BS in Biology from UCLA and her PhD in Zoology from the University of Wisconsin-Madison, where she studied the defensive behavior of neotropical social wasps, and a nesting association between two species of wasps. Over the years, her pets have included dogs, cats, fish, frogs, cockroaches, geckos, ants, anoles, spiders, tarantulas, and a hamster.

She writes for TheWildest.com, writes the animal column for the *Arizona Daily Sun,* and is an Adjunct Professor in the Department of Biological Sciences at Northern Arizona University, where she has enjoyed teaching tropical ecology and conservation field courses in Nicaragua and in Costa Rica. She also teaches a class for freshmen about the importance of insects to society called "Sex, Bugs, and Rock 'n' Roll."

Karen and her husband live in Flagstaff, Arizona, where they raised their two sons.

Instagram: @Karen.London.Dog.Behavior

Made in the USA
Middletown, DE
06 July 2023